Genetic and Evolutionary Computation

Series Editors:
David E. Goldberg, ThreeJoy Associates, Inc., Urbana, IL, USA
John R. Koza, Stanford University, Los Altos, CA, USA

More information about this series at http://www.springer.com/series/7373

Wolfgang Banzhaf • Randal S. Olson
William Tozier • Rick Riolo
Editors

Genetic Programming Theory and Practice XV

 Springer

Editors
Wolfgang Banzhaf
BEACON Center for the Study
of Evolution in Action and
Department of Computer Science
Michigan State University
East Lansing, MI, USA

William Tozier
Institute for Biomedical Informatics
University of Pennsylvania
Philadelphia, PA, USA

Randal S. Olson
Institute for Biomedical Informatics
University of Pennsylvania
Philadelphia, PA, USA

Rick Riolo
Center for the Study of Complex Systems
University of Michigan
Ann Arbor, MI, USA

ISSN 1932-0167
Genetic and Evolutionary Computation
ISBN 978-3-030-08031-0   ISBN 978-3-319-90512-9   (eBook)
https://doi.org/10.1007/978-3-319-90512-9

Printed on acid-free paper

This Springer imprint is published by the registered company Springer International Publishing AG part of Springer Nature.
The registered company address is: Gewerbestrasse 11, 6330 Cham, Switzerland

Foreword

It was a great pleasure giving our keynotes at GPTP 2017. We thank the organizers for inviting us. One of us (Jeff Clune) had always heard of and been interested in attending GPTP, and the other of us (Ken Stanley) had participated before and looked forward to returning. We both greatly enjoyed the workshop. The gathering is unique in its size and format, enabling longer, deeper conversations about the ideas presented than most conferences or workshops allow for. It was exciting to both see many senior leaders in the field, including many old friends, and meet the new researchers who have recently entered the field and are doing great work.

For one of us (Jeff), the setting of this workshop was personally meaningful. I attended the University of Michigan as an undergraduate and had my first taste of research working with Carl Simon, who is the founding director of the Center for the Study of Complex Systems, which is the official host of GPTP. Additionally, when looking for the best university at which to conduct my PhD research, I met with Rick Riolo, long-time GPTP organizer, in the very building GPTP was held in this year. It was great to return to Ann Arbor and that specific building so many years later, thinking back to my origins as a researcher and my desire to dedicate my career to exactly the questions we collectively discussed and study. We are all fortunate to be able to spend our lives researching such fascinating topics. It was especially nice to have Carl visit with the group and for me to be able to personally thank him for kick-starting my research career so many years ago. It is unfortunate Rick could not attend, but it was heartwarming to hear how much of an impact he has had on the GPTP community and how appreciated he is personally and professionally by all GPTP attendees.

Many of the ideas that were presented and discussed were exciting and innovative and have great potential. One that strongly resonated with both of us is Lexicase selection by Lee Spector and his students. One of us (Jeff), along with his PhD student Joost Huizinga, independently invented a similar idea in the combinatorial multi-objective evolutionary algorithm (CMOEA). A great function of conferences is learning about what others in the field are working on and discussing those ideas. Here was a perfect example. Prior to GPTP, Joost and I were not aware of Lexicase selection, but the format of the workshop both allowed us to learn about it during

Lee's presentation and afforded time for a prolonged discussion of it and how it relates to, and differs from, CMOEA. Joost and I have since experimented with Lexicase selection and have added it as a comparison algorithm to our upcoming paper on CMOEA, showing one immediate impact on research the workshop has already had. More generally, both Ken and I love the ideas behind Lexicase and CMOEA and believe they can propel the field toward producing more robust, generalist solutions that solve not just one problem, but many.

There were a multitude of interesting ideas presented and discussed, and we do not have space here to list them all. We will instead quickly mention a few more that particularly resonated with us. One is the work by Randy Olson and Jason Moore on AutoML. We think it is high time to finally realize the long-standing goal of automating machine learning pipelines. Doing so will both expand the impact of machine learning throughout society and catalyze faster progress in machine learning research. A second idea that resonated with us is the Eco-EA algorithm that Charles Ofria presented on. Jeff recalls the original version of that algorithm, which was developed by Sherri Goings, Charles, and collaborators at Michigan State while he was a PhD student. The idea tries to abstract one driver of natural diversity, which is having multiple niches, each with limited resources. If too many agents are currently exploiting one niche (e.g., solving one problem, or solving it in a particular way), there is a reward for some agents in the population to become different in order to exploit less depleted resources.

In fact, we believe that inventing open-ended evolution, in which a computational process endlessly creates an increasingly large set of diverse, high-quality solutions, is one of the great open scientific challenges. We also believe GPTP and similar evolutionary algorithm communities have the potential to make major advances toward this goal. A substantial amount of our own research, including much of what we both presented on in our keynotes, is dedicated to this quest. The Eco-EA algorithm has inspired some of this work, and we are delighted to see that it is continuing to be investigated and enhanced. We would be particularly excited to see how it can be improved to automatically allow niches to be created in a truly open-ended way. Furthermore, at GPTP, some of the best parts of the gathering include the side discussions, and open-endedness was a big topic in those conversations. For Ken, engaging with attendees about some of the hard questions in open-endedness genuinely broadened his appreciation for and understanding of the problem, even after studying it for years. That is the kind of outcome that an intimate and extended gathering like GPTP can offer that other more conventional venues rarely reproduce. We both hope that GPTP can become a catalyst for the growth of open-endedness as a field and community.

One of the discussions at the workshop was how to emulate the success of the community that trains deep neural networks via deep learning. One of us (Jeff) suggested then that what caused the world to take notice of, celebrate, and heavily invest in deep learning is simply that it works extremely well on hard problems. He quoted Steve Martin, who says "be so good they can't ignore you." One example of that was presented at GPTP by Michael Korns. His hedge fund, Korns Associates, built software for investing based on genetic programming, and Korns credited

GPTP as crucial in the development of this software. Korns Associates sold some rights to use this software to Lantern Credit for cash and shares valued at $4.5 million USD. That is a great success story for the community and an example of building something that works so well that it cannot be ignored. Many of the ideas above, and the others presented at the workshop and described in these proceedings, also have the potential to deliver impressive, impossible-to-ignore results. Now the hard work begins to show their true potential. That requires hard science, which inevitably includes work on diagnostic (aka "toy") problems. However, and importantly, it also requires that we increasingly shoot for the stars. That means solving problems so challenging that the world will be forced to take notice of the wonderful work being done by this small but dedicated community. On that note, let's roll up our sleeves, set our ambitions high, and get to work!

Loy and Edith Harris Associate Professor in Computer Science   Jeff Clune
Director, Evolving Artificial Intelligence Lab
University of Wyoming
Senior Research Scientist, Uber AI Labs
San Francisco, CA, USA
Professor in Computer Science   Kenneth O. Stanley
Director, Evolutionary Complexity (EPlex) Lab
University of Central Florida
Senior Research Scientist, Uber AI Labs
San Francisco, CA, USA
February 2018

Preface

The book you hold in hand is the proceedings of the Fifteenth Workshop on Genetic Programming Theory and Practice, an invitation-only workshop held from May 18–20, 2017, at the University of Michigan in Ann Arbor, MI, under the auspices of the Center for the Study of Complex Systems. Since 2003, this annual workshop has been a forum for theorists in and practical users of genetic programming to exchange ideas, share insights, and trade observations. The workshop is intentionally organized as an event, where speculation is welcome and where participants are encouraged to discuss ideas or results that are not necessarily ready for publication in peer-reviewed publication, or have been published in different places and are summarized in the contributions provided for presentation here.

In addition to our regular sessions and interspersed with discussion sessions were three invited keynote talks. While regular talks are usually 40 min to present ideas and take questions, keynote talks are 60 min presentation time plus 10 min for an immediate question and answer session. Often, the ideas of keynote talks provide the start for more in-depth discussions during our discussion sessions.

This year, the first keynote speaker was Jeff Clune, University of Wyoming and Uber AI Lab, with "A talk in two parts: AI Neuroscience, and Harnessing Illumination Algorithms." This talk presented what is at the forefront of what modern AI has to offer these days, mostly through the deep learning technology: very efficient pattern recognition algorithms, in neural network type representations, which are not directly conducive to forming an understanding of what is actually going on. Jeff and his collaborators have come up with a method to examine these deep neural networks to tease out what they have learned in a particular domain, so as to understand and be able to predict what would happen if other patterns were fed into those networks.

The second keynote talk was by Kenneth Stanley of the University of Central Florida and Uber AI on "New Directions in Open-Ended Evolution." Ken, who presently heads the AI lab of Uber, discussed his work on novelty search, and how the benefits of moving away from a simple fitness goal could inform genetic programming and allow it to come up with more creative solutions to problems by promoting diversity through fostering behavioral novelty. His most recent work on minimal criterion evolution featured prominently in his talk and provided plenty of fodder for discussions on open-ended evolution.

The third keynote talk was presented by Patrick Shafto from the Department of Mathematics and Computer Science at Rutgers University. His talk, entitled "Cooperative Inference in Humans and Machines," addressed a very important new development in AI/ML—the collaboration of humans and computers to extract information and produce knowledge from data. It turns out that in the age of big data this cooperation is much more efficient in producing valuable insights than either computer algorithms or human learning.

We hope that the contributions published in this collection provide an exciting snapshot of what is going on in genetic programming!

Acknowledgements

We would like to thank all of the participants for again making GP Theory and Practice a successful workshop 2017. As is always the case, it produced a lot of interesting and high-energy discussions, as well as speculative thoughts and new ideas for further work. The keynote speakers did an excellent job at raising our awareness and provided thought-provoking ideas about the potential of genetic programming and its place in the world.

We would also like to thank our financial supporters for making the existence of GP Theory and Practice possible for the past 15 years, and counting. For 2017, these include:

- The Center for the Study of Complex Systems at the University of Michigan, and especially Carl Simon and Charles Doering, the champions of the workshop series
- John Koza
- Michael F. Korns, Lantern LLC
- Stuart W. Card
- Thomas Kern

A number of people made key contributions to the organization and assisted our participants during their stay in Ann Arbor. Foremost among them are Linda Wood and Mita Gibson who made the workshop run smoothly with their diligent efforts behind the scene before, during, and after the workshop. Special thanks go to the Springer Publishing Company, for providing the editorial assistance for producing this book. We are particularly grateful for contractual assistance by Melissa Fearon at Springer and all their staff has done to make this book possible.

East Lansing, MI, USA	Wolfgang Banzhaf
Philadelphia, PA, USA	Randall S. Olson
Ann Arbor, MI, USA	William Tozier
Ann Arbor, MI, USA	Rick Riolo
February 2018	

Contents

Contributors

Michael Affenzeller Heuristic and Evolutionary Algorithms Laboratory, University of Applied Sciences Upper Austria, Hagenberg, Austria

Institute for Formal Models and Verification, Johannes Kepler University, Linz, Austria

Wolfgang Banzhaf BEACON Center for the Study of Evolution in Action and Department of Computer Science, Michigan State University, East Lansing, MI, USA

Bogdan Burlacu Heuristic and Evolutionary Algorithms Laboratory, University of Applied Sciences Upper Austria, Hagenberg, Austria

Institute for Formal Models and Verification, Johannes Kepler University, Linz, Austria

Suart W. Card Syracuse University, Syracuse, NY, USA

Emily Dolson BEACON Center for the Study of Evolution in Action and Department of Computer Science and Ecology, Evolutionary Biology, and Behavior Program, Michigan State University, East Lansing, MI, USA

Steven B. Fine MIT CSAIL, Cambridge, MA, USA

Weixuan Fu Institute for Biomedical Informatics, University of Pennsylvania, Philadelphia, PA, USA

Amir H. Gandomi School of Business, Stevens Institute of Technology, Hoboken, NJ, USA

BEACON Center for the Study of Evolution in Action, Michigan State University, East Lansing, MI, USA

Thomas Helmuth Hamilton College, Clinton, NY, USA

Erik Hemberg MIT CSAIL, Cambridge, MA, USA

John H. Holmes Institute for Biomedical Informatics, University of Pennsylvania, Philadelphia, PA, USA

Domagoj Jakobovic University of Zagreb, Faculty of Electrical Engineering and Computing, Zagreb, Croatia

Michael Kommenda Heuristic and Evolutionary Algorithms Laboratory, University of Applied Sciences Upper Austria, Hagenberg, Austria

Institute for Formal Models and Verification, Johannes Kepler University, Linz, Austria

Michael F. Korns Lantern Credit LLC, Henderson, NV, USA

Krzysztof Krawiec Poznan Institute of Technology, Poznań, Poland

Gabriel Kronberger Heuristic and Evolutionary Algorithms Laboratory, University of Applied Sciences Upper Austria, Hagenberg, Austria

William La Cava Institute for Biomedical Informatics, University of Pennsylvania, Philadelphia, PA, USA

Kishan G. Mehrotra Syracuse University, Syracuse, NY, USA

Chilukuri K. Mohan Syracuse University, Syracuse, NY, USA

Jason H. Moore Institute for Biomedical Informatics, University of Pennsylvania, Philadelphia, PA, USA

Charles Ofria BEACON Center for the Study of Evolution in Action and Department of Computer Science and Ecology, Evolutionary Biology, and Behavior Program, Michigan State University, East Lansing, MI, USA

Randal S. Olson Institute for Biomedical Informatics, University of Pennsylvania, Philadelphia, PA, USA

Una-May O'Reilly MIT CSAIL, Cambridge, MA, USA

Patryk Orzechowski Institute for Biomedical Informatics, University of Pennsylvania, Philadelphia, PA, USA

Department of Automatics and Biomedical Engineering, AGH University of Science and Technology, Krakow, Poland

Edward Pantridge MassMutual, Amherst, MA, USA

Stjepan Picek MIT CSAIL, Cambridge, MA, USA

Saul Shanabrook University of Massachusetts, Amherst, MA, USA

Moshe Sipper Institute for Biomedical Informatics, University of Pennsylvania, Philadelphia, PA, USA

Department of Computer Science, Ben-Gurion University, Beer-Sheva, Israel

Lee Spector Hampshire College, Amherst, MA, USA

Amirhessam Tahmassebi Department of Scientific Computing, Florida State University, Tallahassee, FL, USA

Sharon Tartarone Institute for Biomedical Informatics, University of Pennsylvania, Philadelphia, PA, USA

Sarah Anne Troise Washington and Lee University, Lexington, VA, USA

Ryan J. Urbanowicz Institute for Biomedical Informatics, University of Pennsylvania, Philadelphia, PA, USA

Steven Vitale Institute for Biomedical Informatics, University of Pennsylvania, Philadelphia, PA, USA

Stephan Winkler Heuristic and Evolutionary Algorithms Laboratory, University of Applied Sciences Upper Austria, Hagenberg, Austria

Institute for Formal Models and Verification, Johannes Kepler University, Linz, Austria

Ben Yang Institute for Biomedical Informatics, University of Pennsylvania, Philadelphia, PA, USA

Zhiruo Zhao Syracuse University, Syracuse, NY, USA

Chapter 1
Exploiting Subprograms in Genetic Programming

Steven B. Fine, Erik Hemberg, Krzysztof Krawiec, and Una-May O'Reilly

Abstract Compelled by the importance of subprogram behavior, we investigate how much Behavioral Genetic Programming is sensitive to model bias. We experimentally compare two different decision tree algorithms analyzing whether it is possible to see significant performance differences given that the model techniques select different subprograms and differ in how accurately they can regress subprogram behavior on desired outputs. We find no remarkable difference between REPTree and CART in this regard, though for a modest fraction of our datasets we find that one algorithm results in superior error reduction than the other. We also investigate alternative ways to identify useful subprograms beyond examining those within one program. We propose a means of identifying subprograms from different programs that work well together. This method combines behavioral traces from multiple programs and uses the information derived from modeling the combined program traces.

1.1 Introduction

Our general goal is to improve the level of program complexity that genetic programming (GP) can routinely evolve [7]. This is toward fulfilling its potential to occupy a significant niche in the ever advancing field of program synthesis [4, 15, 19]. Behavioral genetic programming (BGP) is an extension to GP that advances toward this compelling goal [8, 9]. The intuition of the BGP paradigm is that during evolutionary search and optimization, we can identify information characterizing programs by behavioral properties that extend beyond how accurately

S. B. Fine · E. Hemberg · U.-M. O'Reilly (✉)
MIT CSAIL, Cambridge, MA, USA
e-mail: sfine@mit.edu; hembergerik@csail.mit.edu; unamay@csail.mit.edu

K. Krawiec
Poznan Institute of Technology, Poznań, Poland
e-mail: krzysztof.krawiec@cs.put.poznan.pl

© Springer International Publishing AG, part of Springer Nature 2018
W. Banzhaf et al. (eds.), *Genetic Programming Theory and Practice XV*,
Genetic and Evolutionary Computation, https://doi.org/10.1007/978-3-319-90512-9_1

they match their target outputs. This information from a program "trace" can be effectively integrated into extensions of the algorithm's fundamental mechanisms of fitness-based selection and genetic variation.

To identify useful subprograms BGP commonly exploits the program *trace* information (first introduced in [10] which is a capture of the output of every subprogram within the program for every test data point during fitness evaluation).

The trace is stored in a matrix where the number of rows is equal to the number of test suite data points, and the number of columns is equal to the number of subtrees in a given program. The trace captures a full snapshot of all of the intermediate states of the program evaluation.

BGP then uses the trace to estimate the merit of each subprogram by treating each column as a feature (or explanatory variable) in a *model regression* on the desired program outputs. The accuracy and complexity of the model reveals how useful the subprograms are. If the model has feature selection capability, it also reveals specific subprograms within the tree that are partially contributing to the program's fitness. BGP uses this information in two ways. It integrates *model error* and *model complexity* into the program's fitness . Second, it maintains an archive of the most useful subtrees identified by modeling each program and uses them in an *archive-based crossover* . BGP has a number of variants that collectively yield impressive results, see [8, 9].

In this work, we explore various extensions to the BGP paradigm that are motivated by two central topics:

1. **The impact of bias from the inference model on useful subprogram identi-fication and program fitness.** Model techniques and even the implementation of the same technique can differ in inductive *bias*, i.e. error from assumptions in the learning algorithm, e.g. implementation of a "decision tree" algorithm. These differences, in turn, impact which subprograms are inserted/retrieved from the BGP archive and the model accuracy and model complexity factors that are integrated into a program's fitness. Therefore, we investigate how sensitive BGP is to model bias.

   How important is it which subprograms the model technique selects and how accurate a model is? We answer these questions by comparing BGP competence under two different implementations of decision tree modeling which we observe have different biases. Our investigation contrasts feature identification and model accuracy under the two implementations.

2. **"Does it play well with others?"**: **Alternate ways to identify useful sub-programs**. In BGP, the trace matrix is calculated for each program in the population. This means that feature selection and subprogram fitness estimation occur within the program context. Essentially, each subprogram is juxtaposed with "relatives"—its parent, neighbors and even children in GP tree terms. Does this context provide the best means of identifying useful subprograms? It may not. Crossover moves a subprogram into another program so, to work most effectively, the BGP process should explore recombinations of subprograms that *work well with other subprograms and programs in the population.* We examine

this idea by concatenating program traces from a set of programs, not solely one program. We then feed the concatenated trace into a model regression. This demands a new measure of fitness to reflect how many subprograms each program contributes to the resulting model. This new measure of fitness can be integrated into the program's fitness. We examine concatenation of the entire population and sub-populations that are selected based on fitness.

Our specific demonstration focus herein is symbolic regression (SR). We choose SR because it remains a challenge and has good benchmarks [12] so it allows us to measure progress and extensions. It also has real world application to system identification and, with modest modification, machine learning regression and classification. In passing, we replicate BGP logic, making our project software available with an open source license.

We proceed as follows. We start with related work in Sect. 1.2. In Sect. 1.3 we provide descriptions of our methods of comparison and investigation. In Sect. 1.4 we provide implementation and experimental details and results. Section 1.5 concludes and mentions future work.

1.2 Related Work

BGP is among a number of other approaches to program synthesis where progress has recently become more empirically driven, rather than driven by formal specifications and verification [1]. Alternative approaches to evolutionary algorithms include sketching [16], i.e. communicating insight through a partial program, generalized program verification [17], as well as hybrid computing with neural network and external memory [3].

BGP takes inspiration, with respect to its focus on program behavior, from earlier work on implicit fitness sharing [13], trace consistency analysis and the use of archives as a form of memory [6]. In its introduction in [10] the preceding introduction of the trace matrix was noted within a system called Pattern Guided Genetic Programming, "PANGEA". In PANGEA minimum description length was induced from the program execution. Subsequently there have been a variety of extensions. For example, in the general vein of behaviorally characterizing a program by more than its accuracy [11] considers discovering search objectives for test-based problems. Also notable is Memetic Semantic Genetic Programming [2].

BGP was introduced as a broader and more detailed take on Semantic GP. BGP and Semantic GP share the common goal of characterising program behavior in more detail and exploiting this information. Whereas BGP looks at subprograms, semantic GP focuses on program output. Output is scrutinized for every test individually and a study of the impact of crossover on program semantics and semantic building blocks [14] was conducted. A survey of semantic models in GP [18] gives an overview of methods for their operators, as well as different objectives.

1.3   Exploiting Subprograms

1.3.1   BGP Strategy

What emerges from the details of BGP's successful examples is a stepwise strategy:

1. For each program in the population, capture the behavior of each subprograms in a trace matrix T.
2. Regress T as feature data on the desired program outputs and derive a model M.
3. Assign a value of merit w to each subprogram in M. Use this merit to determine whether it should be inserted into an archive. Use a modified crossover that draws subprograms from the archive.
4. Integrate model error e and complexity c into program fitness.

One example of the strategy is realized in [9] where, in Step (2), the fast RepTree (REPT—Reduced Error Pruning Tree) algorithm of decision tree classification from the WEKA Data Mining software library [5] is used for regression modeling. REPT builds a decision/regression tree using information gain/variance. In Step (3) merit is measured per Eq. (1.1) where $|U(p)|$ is the number of subprograms (equivalently distinct columns of the trace) used in the model and e is model error.

$$w = \frac{1}{(1 + e)|U(p)|} \tag{1.1}$$

1.3.2   Exploring Model Bias

Following our motivation to understand the impact of model bias on useful subprogram identification and program fitness, we first explore an alternative realization of BGP's strategy by using the CART optimized version of the CART decision tree algorithm.[1] CART (Classification and Regression Trees) is very similar to C4.5, but it differs in that it supports numerical target variables (regression) and does not compute rule sets. CART constructs binary trees using the feature and threshold that yield the largest information gain at each node. With the CART implementation we derive a model in the same manner that we derive a model from REPT. We denote the model derived for CART by M_S and contrast it with the model derived from REPT, which we now denote by M_R.

---

[1] http://scikit-learn.org/stable/modules/tree.html#tree-algorithms-id3-c4-5-c5-0-and-cart.

1.3.3   Identifying Useful Subprograms

Next we realize alternative implementations of the BGP strategy. We do this to investigate alternative ways of identifying useful subprograms, considering that prior work only considers models that are trained on the trace of a single program. In Step (1) we first select a set of programs C from the population. We then form a new kind of trace matrix, T_c, by column-wise concatenating all T's of the programs in C. In a version we call **FULL**, T_c is then passed through Step (2). We proceed with Step (3), but it is important to note that the weights given to the subprograms considered for the archive are identical, because only a single model is built. Step (4) is altered to incorporate model contribution in place of model error and complexity. The model is built from the features of many programs, so the model error and model complexity for each individual program are undefined. Model contribution measures how many features each program contributes to M. For this we use w_c, which is given by Eq. (1.2), where p' is the number of features in M from program p, and $|U(M)|$ is the total number of features from T_c used in M. This method allows us to experiment with different programs in C, trying C containing all the programs in the population, for diversity and, conversely trying elitism, holding only a top fraction of the population by fitness.

$$w_c = 1 - \frac{p'}{|U(M)|} \tag{1.2}$$

An alternative implementation, that we name **DRAW**, is to draw random subsets from T_c, and build a model on each. This would possibly contribute to a more robust archive, if it can be populated with subtrees that are frequently selected by the machine learning model. We modify Step (3) to account for the possibility that a specific subtree was chosen to be in more than one model. In this case, the subtree's merit w is set to be the sum of the assigned merit values each time the subtree is chosen. Step (4) is modified in the same way as it is in **FULL**.

Implementation details of **FULL** and of **DRAW** are provided in the next section.

1.4   Experiments

We start this section by detailing the benchmarks we use, the parameters of our algorithms and name our algorithm configurations for convenience. Section 1.4.2 then evaluates the impact of different decision tree algorithms. Section 1.4.3 evaluates the performance of **FULL** and **DRAW** for different configurations then compares **FULL**, **DRAW** and program-based model techniques.

1.4.1 Experimental Data, Parameters

Our investigation uses 17 symbolic regression benchmarks from [12]. All of the benchmarks are defined such that the dependent variable is the output of a particular mathematical function for a given set of inputs. All of the inputs are taken to form a grid on some interval. Let $E[a, b, c]$ denote c samples equally spaced in the interval $[a, b]$. (Note that McDermott et al. defines $E[a, b, c]$ slightly differently.) Below is a list of all of the benchmarks that are used:

1. **Keijzer1**: $0.3x \sin(2\pi x)$; $x \in E[-1, 1, 20]$
2. **Keijzer11**: $xy + \sin((x - 1)(y - 1))$; $x, y \in E[-3, 3, 5]$
3. **Keijzer12**: $x^4 - x^3 + \frac{y^2}{2} - y$; $x, y \in E[-3, 3, 5]$
4. **Keijzer13**: $6 \sin(x) \cos(y)$; $x, y \in E[-3, 3, 5]$
5. **Keijzer14**: $\frac{8}{2+x^2+y^2}$; $x, y \in E[-3, 3, 5]$
6. **Keijzer15**: $\frac{x^3}{5} - \frac{y^3}{2} - y - x$; $x, y \in E[-3, 3, 5]$
7. **Keijzer4**: $x^3 e^{-x} \cos(x) \sin(x)(\sin^2(x)\cos(x) - 1)$; $x \in E[0, 10, 20]$
8. **Keijzer5**: $\frac{3xz}{(x-10)y^2}$; $x, y \in E[-1, 1, 4]$; $z \in E[1, 2, 4]$
9. **Nguyen10**: $2 \sin(x) \cos(y)$; $x, y \in E[0, 1, 5]$
10. **Nguyen12**: $x^4 - x^3 + \frac{y^2}{2} - y$; $x, y \in E[0, 1, 5]$
11. **Nguyen3**: $x^5 + x^4 + x^3 + x^2 + x$; $x \in E[-1, 1, 20]$
12. **Nguyen4**: $x^6 + x^5 + x^4 + x^3 + x^2 + x$; $x \in E[-1, 1, 20]$
13. **Nguyen5**: $\sin(x^2) \cos(x) - 1$; $x \in E[-1, 1, 20]$
14. **Nguyen6**: $\sin(x) + \sin(x + x^2)$; $x \in E[-1, 1, 20]$
15. **Nguyen7**: $\ln(x + 1) + \ln(x^2 + 1)$; $x \in E[0, 2, 20]$
16. **Nguyen9**: $\sin(x) + \sin(y^2)$; $x, y \in E[0, 1, 5]$
17. **Sext**: $x^6 - 2x^4 + x^2$; $x \in E[-1, 1, 20]$

We use a standard implementation of GP and chose parameters according to settings documented in [9].

Fixed Parameters

- **Tournament size**: 4
- **Population size**: 100
- **Number of Generations**: 250
- **Maximum Program Tree Depth**: 17
- **Function set**[2]: $\{+, -, *, /, \log, \exp, \sin, \cos, -x\}$
- **Terminal set**: Only the features in the benchmark.
- **Archive Capacity**: 50
- **Mutation Rate** μ: 0.1
- **Crossover Rate with Archive configuration** χ: 0.0

---

[2]Note that for our implementation of $/$, if the denominator is less than 10^{-6} we return 1, and for our implementation of log, if the argument is less than 10^{-6} we return 0.

- **Crossover Rate with GP** χ: 0.9
- **Archive-Based Crossover Rate** α: 0.9
- **REPTree** defaults but no pruning
- **CART** defaults
- **Number of runs** 30

We use the same four program fitness functions used in [9] (in addition to model contribution fitness which is described in Sect. 1.3.3). Program error f, is given by Eq. (1.3), where $\hat{y}$ is the output of the program, y is the desired output, and d_m denotes the Manhattan distance between the two arguments. Program size s, is given by Eq. (1.4), where $|p|$ is the number of nodes in the tree that defines the program. Model error e, is given by Eq. (1.5), where M is the output of the machine learning model when it is evaluated on the trace of the program. Model complexity c, is given by Eq. (1.6), where $|M|$ is the size of the model.

$$f = 1 - \frac{1}{1 + d_m(\hat{y}, y)} \tag{1.3}$$

$$s = 1 - \frac{1}{|p|} \tag{1.4}$$

$$e = 1 - \frac{1}{1 + d_m(M, y)} \tag{1.5}$$

$$c = 1 - \frac{1}{|M|} \tag{1.6}$$

First we use the 3 BGP algorithm configurations that use REPT to replicate [9]'s work on the symbolic regression benchmarks. These we call BP2A, BP4, BP4A following precedent. In the name the digit 2 indicates that model error e and complexity c were not integrated into program fitness while 4 indicates they were. The suffix A indicates whether or not subprograms from the model were qualified for archive insertion and archive retrieval during BGP crossover. When the A is omitted ordinary crossover is used. We observe results consistent with the prior work. Our open source software is available on Github.[3] This allowed us to proceed to evaluate feature selection sensitivity to the modeling algorithm.

It is important to note, that for each configuration we report regression, i.e. training set performance. We are primarily interested in exploring subprogram behavior and how to assemble subprograms. Reporting generalization would complicate the discussion without materially affecting our conclusions.

---

[3]https://github.com/flexgp/BehavioralGP.

Table 1.1 Comparison of
impact of REPT vs CART for
average fitness rank across all
data sets

	Configuration		Average rank
1	BP2A	REPT	1.82
2	BP2A	CART	2.94
3	BP4A	CART	3.06
4	BP4	CART	3.18
5	BP4A	REPT	4.65
6	BP4	REPT	5.35

1.4.2   Sensitivity to Model Bias

Q1. Does the feature selection bias of the model step matter?

Table 1.1 shows the results of running the three different configurations (BP2A, BP4, BP4A) each with the two decision tree algorithms. Averaging over the rankings across each benchmark we find that BP2A using REPT is best. For BP2A, REPT outranks CART but when model error is integrated into the program fitness, (i.e. BP4A and BP4) regardless of whether or not an archive is used, CART is superior to REPT.

When we compare the results of using the archive while model error is integrated into the program fitness (i.e. BP4A to BP4), for both REPT and CART it is better to use an archive than to forgo one. Comparing BP2A with BP4A, we can measure the impact of model error and complexity integration. We find that for both CART and REPT it is not helpful to integrate model error and complexity into program fitness.

For a deeper dive, at the specific benchmark level, Table 1.2 shows the average best fitness at end of run (of 30 runs), for each benchmark. Averaging all fitness results, no clear winner is discernible. For certain comparisons CART will be superior while for others REPT is. We also show one randomly selected run of Keijzer1 running with REPT modeling and configuration BP4 in Fig. 1.1. We plot on the first row model error on the left and the fitness of the best program (right). The plots on the second row show number of features of model and number of subprograms in the best program (right). The plots on the third row show the ratio of number of model features to program subtrees (left) and ratio of model error to program fitness. Since the run is configured for BP4 program fitness integrates both model error and complexity. No discernible difference arose among this sort of plot. This is understandable given the stochastic nature of BGP.

We conclude that in this case of different decision tree algorithms perhaps the subtlety of contrast is not strong enough.

1.4.3   Aggregate Trace Matrices

In this section, we compare various configurations of **FULL** and **DRAW**. For the algorithm configurations of this section, we adopt a clearer notation. We drop

Table 1.2 Comparison of different decision tree algorithms: REPT and CART on average program error for best of run programs (averaged across 30 runs)

		Keij1	Keij11	Keij12	Keij13	Keij14	Keij15	Keij4	Keij5	–
BP2A	REPT	**0.243**	0.776	0.972	**0.393**	0.723	**0.883**	**0.384**	**0.975**	–
	CART	0.327	0.769	**0.966**	0.481	0.726	0.907	0.468	0.977	–
BP4	REPT	0.359	0.852	0.982	0.817	0.872	0.922	0.522	0.993	–
	CART	0.357	**0.684**	0.968	0.548	0.776	0.887	0.513	0.991	–
BP4A	REPT	0.319	0.804	0.981	0.765	0.821	0.919	0.505	0.991	–
	CART	0.261	0.811	0.973	0.507	**0.691**	0.94	0.471	0.981	–

		Nguy10	Nguy12	Nguy3	Nguy4	Nguy5	Nguy6	Nguy7	Nguy9	Sext
BP2A	REPT	**0.11**	**0.343**	0.196	**0.265**	0.037	0.091	0.122	0.068	**0.052**
	CART	0.199	0.379	0.2	0.285	0.04	0.119	0.127	0.075	0.054
BP4	REPT	0.309	0.388	**0.193**	0.33	0.103	0.133	0.117	0.165	0.127
	CART	0.144	0.36	0.266	0.288	0.126	**0.0**	**0.104**	**0.04**	0.083
BP4A	REPT	0.209	0.386	0.22	0.328	0.088	0.117	0.128	0.194	0.1
	CART	0.264	0.379	0.219	0.273	**0.034**	0.088	0.115	0.065	0.056

N.B. program error does NOT include program size. During evolution the fitness of a program integrates program error and size per [9]

the BGP prefix and use M to denote when program contribution is integrated into program fitness, and $\hat{M}$ to denote when it is not. We use A to denote when subprograms are qualified for archive insertion and archive retrieval during BGP crossover, and $\hat{A}$ to denote when ordinary crossover is used.

More details of the **DRAW** method are appropriate. Referencing [9] we analyze the formula for computing the weight of a given subtree (see Eq. (1.1)). We note that the $|U(p)|$ factor in its denominator indirectly increases the weight of smaller subprograms. This occurs because smaller programs yield smaller models (i.e. smaller $|U(p)|$), and smaller programs have smaller subprograms. Therefore we designed **DRAW** to also favor the archiving of smaller subprograms. **DRAW** proceeds as follows:

1. The population is sorted best to worst by program fitness (program error and size) using the NSGA pareto front crowding calculation because BGP is multi-objective.
2. The sorted population is cut off from below at a threshold $\lambda\%$ to form C. The trace matrixes of every program in C are concatenated to form T_C which we call the subprogram pool.
3. We next sort the population by **size** and select the smallest 20% forming a size sample we call K.
4. Finally we draw from K at random to obtain the number of subprograms that will be collectively modeled. Then we select the equivalent number of columns at random from T_C and form a model. We repeat this step each time for the size of the population. This generates multiple smaller collections of diverse subprograms.

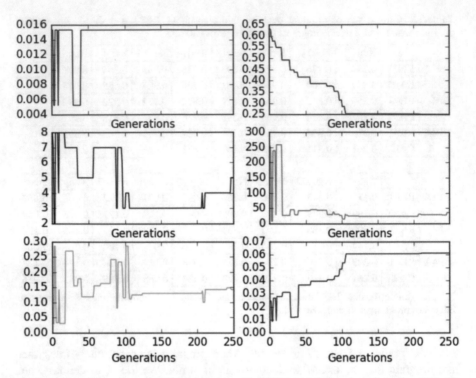

Fig. 1.1 We take one run of Keijzer1 running with REPT modeling and configuration BP4. We plot on the first row model error on the left and the fitness of the best program (right). The plots on the second row show number of features of model and number of subprograms in the best program (right). The plots on the third row show the ratio of number of model features to program subtrees (left) and ratio of model error to program fitness. Since the run is configured for BP4 program fitness integrates both model error and complexity

Q2. Can trace matrix concatenation which pools subprograms among different programs improve BGP performance?

We first asked what if C is composed of **every** subprogram in the population, i.e. $|C| = PopSize$? While this C using **FULL** would only support one model being derived, it would give all subprograms in the population an opportunity to be used with each other in the model as features. Similarly, by favoring many smaller combinations drawn from all subprograms, **DRAW** would, through repetition, give all subprograms in the population an opportunity to be used with some of the all the others. If we compare the result of **DRAW** and **FULL** we can gauge the difference between generating many more small models vs one bigger model, when every subprogram in the population is "eligible" to be selected as a model feature. This comparison is detailed on the bottom line of Table 1.3. The leftmost averaged ranking results (by average fitness, across the 17 benchmarks) for different model and archive options are from **DRAW** and the rightmost are from **FULL**. The data reveal that using all the subprograms, with either **FULL** or **DRAW** is

Table 1.3 DRAW (lhs) and
FULL (rhs) average rank
varying model fitness signal
(M or $\hat{M}$) and use of archive
(A or $\hat{A}$) for 17 benchmarks

C	$\hat{M}A$	$M\hat{A}$	MA	$\hat{M}A$	$M\hat{A}$	MA
25	3.06	2.24	2.35	**1.65**	2.18	**1.41**
50	**2.29**	**1.82**	1.88	2.41	**1.88**	2.29
75	**2.29**	2.0	2.0	3.06	2.12	2.65
100	2.35	3.94	3.76	2.88	3.82	3.65

NOT advantageous. Further empirical investigation to understand this result should consider two issues: (1) the program size to fitness distribution of the population each generation could be leading to very large number of subprograms and (2) the modeling algorithm (REPT) may be overwhelmed, in the case of **FULL**, by the number of features, given the much smaller number of training cases for the regression.

Next we can consider the rankings of each configuration across different selections for the subprogram pool C. When $\lambda = 25$ the model feature options are from the highest fitness tier of the population. In four of six cases, this appears to *impede* the error of the best of run program, as measured by average ranking. In four of six cases, including all three of **DRAW**, sizing the subprogram pool to be slightly less elitist ($\lambda = 50$ or $\lambda = 75$) was better. But extending λ to 100 appears to be too diverse. Tables 1.4 and 1.5 provide more detailed average fitness and ranking information, i.e. results for each individual benchmark.

Finally, we compare these configurations to the three original BGP configurations. We find that the best performing method is highly dependent on the specific benchmark, and that overall none of the configurations is shown to be the clear winner.

1.5 Conclusions and Future Work

The paper's primary contributions are to explore two subprogram value driven questions. The first question addressed the importance of a choice of modeling algorithm. The modeling algorithm can impact selective pressure (because model fitness can be integrated back into program fitness) and genetic variation (because subprograms used by a model can be inserted into the BGP archive and used in BGP crossover). We tried two algorithms for decision trees: REPT and CART. Neither of the algorithms produced significantly better results across all the 17 benchmarks. For some benchmarks the average fitness results were significantly different but, again, neither of algorithms was consistently superior in each case. Using a completely different modeling technique, i.e. one different from decision trees altogether, that also provides feature selection would be an interesting comparison to using REPT. All feature selection algorithms are stochastic so their results vary. Perhaps the stochasticity of any algorithm overwhelms the impact of a particular technique's bias. We next will consider whether the stronger dissimilarity

Table 1.4 Sampling subprograms for modeling across the population, not from one program

		Keij1	Keij11	Keij12	Keij13	Keij14	Keij15	Keij4	Keij5	–
$\hat{M}A$	Draw 25	0.306	0.704	0.976	0.436	0.768	**0.845**	0.371	0.975	–
	Draw 50	0.286	**0.604**	0.969	0.422	0.731	0.866	0.376	**0.967**	–
	Draw 75	**0.246**	0.716	0.968	**0.325**	0.736	0.869	0.347	0.974	–
	Draw 100	0.253	0.695	0.969	0.382	**0.716**	0.877	**0.33**	0.972	–
	Full 25	0.278	0.812	**0.956**	0.621	0.761	0.88	0.457	0.977	–
	Full 50	0.279	0.883	0.979	0.564	0.748	0.921	0.411	0.981	–
	Full 75	0.302	0.864	0.976	0.604	0.804	0.925	0.453	0.982	–
	Full 100	0.272	0.864	0.982	0.565	0.809	0.947	0.397	0.977	–
$M\hat{A}$	Draw 25	0.322	0.89	0.979	0.732	0.786	0.889	0.601	0.991	–
	Draw 50	0.314	0.824	0.979	0.697	0.798	0.888	0.55	0.986	–
	Draw 75	0.337	0.865	0.979	0.723	0.819	0.886	0.562	0.99	–
	Draw 100	0.367	0.908	0.986	0.879	0.846	0.962	0.598	0.993	–
	Full 25	0.288	0.875	0.973	0.478	0.783	0.867	0.516	0.987	–
	Full 50	0.301	0.851	0.967	0.463	0.834	0.894	0.52	0.984	–
	Full 75	0.317	0.824	0.974	0.538	0.781	0.886	0.499	0.986	–
	Full 100	0.368	0.833	0.979	0.708	0.838	0.949	0.536	0.991	–
MA	Draw 25	0.301	0.808	0.976	0.625	0.798	0.919	0.385	0.984	–
	Draw 50	0.295	0.803	0.975	0.54	0.735	0.927	0.404	0.984	–
	Draw 75	0.292	0.797	0.975	0.567	0.73	0.937	0.426	0.988	–
	Draw 100	0.306	0.866	0.986	0.814	0.751	0.961	0.489	0.991	–
	Full 25	0.304	0.847	0.974	0.685	0.767	0.936	0.498	0.985	–
	Full 50	0.315	0.872	0.981	0.656	0.763	0.936	0.421	0.988	–
	Full 75	0.317	0.902	0.984	0.626	0.78	0.95	0.474	0.986	–
	Full 100	0.326	0.903	0.987	0.759	0.81	0.953	0.496	0.989	–

		Nguy10	Nguy12	Nguy3	Nguy4	Nguy5	Nguy6	Nguy7	Nguy9	Sext
$\hat{M}A$	Draw 25	0.162	0.341	**0.172**	0.301	0.056	0.074	0.132	0.241	0.058
	Draw 50	0.107	0.353	0.194	0.295	0.045	0.089	**0.103**	0.159	0.059
	Draw 75	**0.089**	0.351	0.215	0.278	0.06	0.1	0.118	0.206	**0.044**
	Draw 100	0.123	0.356	0.217	0.285	**0.03**	0.129	0.115	0.165	0.047
	Full 25	0.232	0.385	0.297	0.33	0.058	0.159	0.204	0.212	0.062
	Full 50	0.294	0.387	0.337	0.37	0.06	0.264	0.195	0.301	0.086
	Full 75	0.364	0.395	0.316	0.361	0.059	0.271	0.225	0.306	0.088
	Full 100	0.304	0.393	0.376	0.372	0.081	0.277	0.179	0.214	0.129
$M\hat{A}$	Draw 25	0.185	0.361	0.233	0.283	0.107	0.103	0.144	0.197	0.076
	Draw 50	0.183	0.393	0.307	0.322	0.081	0.088	0.15	0.165	0.081
	Draw 75	0.22	0.356	0.236	0.246	0.064	0.108	0.136	0.242	0.088
	Draw 100	0.377	0.43	0.363	0.442	0.156	0.186	0.246	0.267	0.127
	Full 25	0.163	0.346	0.23	0.302	0.072	**0.069**	0.119	0.174	0.093
	Full 50	0.121	**0.338**	0.23	0.274	0.048	0.117	0.144	**0.132**	0.066
	Full 75	0.168	0.353	0.188	**0.23**	0.067	0.085	0.161	0.172	0.074
	Full 100	0.218	0.384	0.321	0.318	0.083	0.198	0.157	0.24	0.129

(continued)

Table 1.4 (continued)

M A	Draw 25	0.211	0.361	0.287	0.329	0.082	0.205	0.147	0.336	0.072
	Draw 50	0.235	0.349	0.303	0.296	0.073	0.204	0.156	0.287	0.074
	Draw 75	0.274	0.364	0.257	0.329	0.06	0.171	0.124	0.318	0.074
	Draw 100	0.315	0.408	0.393	0.455	0.113	0.255	0.24	0.257	0.093
	Full 25	0.282	0.358	0.295	0.384	0.067	0.262	0.188	0.271	0.086
	Full 50	0.349	0.369	0.356	0.452	0.072	0.302	0.271	0.289	0.098
	Full 75	0.326	0.394	0.397	0.436	0.097	0.351	0.239	0.357	0.13
	Full 100	0.428	0.423	0.507	0.449	0.158	0.383	0.268	0.236	0.168

Two methods **DRAW** and **FULL** were evaluated. Data shows average fitness of each algorithm configuration across all benchmarks

Table 1.5 Rank based program error for best of run programs

		Keij1	Keij11	Keij12	Keij13	Keij14	Keij15	Keij4	Keij5	–
$\hat{M}A$	Draw 25	4	3	4	4	4	1	3	4	–
	Draw 50	3	1	3	3	2	2	4	1	–
	Draw 75	1	4	1	1	3	3	2	3	–
	Draw 100	2	2	2	2	1	4	1	2	–
	Full 25	2	1	1	4	2	1	4	2	–
	Full 50	3	4	3	1	1	2	2	3	–
	Full 75	4	2	2	3	3	3	3	4	–
	Full 100	1	3	4	2	4	4	1	1	–
$M\hat{A}$	Draw 25	2	3	2	3	1	3	4	3	–
	Draw 50	1	1	3	1	2	2	1	1	–
	Draw 75	3	2	1	2	3	1	2	2	–
	Draw 100	4	4	4	4	4	4	3	4	–
	Full 25	1	4	2	2	2	1	2	3	–
	Full 50	2	3	1	1	3	3	3	1	–
	Full 75	3	1	3	3	1	2	1	2	–
	Full 100	4	2	4	4	4	4	4	4	–
M A	Draw 25	3	3	3	3	4	1	1	2	–
	Draw 50	2	2	1	1	2	2	2	1	–
	Draw 75	1	1	2	2	1	3	3	3	–
	Draw 100	4	4	4	4	3	4	4	4	–
	Full 25	1	1	1	3	2	1	4	1	–
	Full 50	2	2	2	2	1	2	1	3	–
	Full 75	3	3	3	1	3	3	2	2	–
	Full 100	4	4	4	4	4	4	3	4	–

(continued)

Table 1.5 (continued)

		Nguy10	Nguy12	Nguy3	Nguy4	Nguy5	Nguy6	Nguy7	Nguy9	Sext
$\hat{M}A$	Draw 25	4	1	1	4	3	1	4	4	3
	Draw 50	2	3	2	3	2	2	1	1	4
	Draw 75	1	2	3	1	4	3	3	3	1
	Draw 100	3	4	4	2	1	4	2	2	2
	Full 25	1	1	1	1	1	1	3	1	1
	Full 50	2	2	3	3	3	2	2	3	2
	Full 75	4	4	2	2	2	3	4	4	3
	Full 100	3	3	4	4	4	4	1	2	4
$M\hat{A}$	Draw 25	2	2	1	2	3	2	2	2	1
	Draw 50	1	3	3	3	2	1	3	1	2
	Draw 75	3	1	2	1	1	3	1	3	3
	Draw 100	4	4	4	4	4	4	4	4	4
	Full 25	2	2	2	3	3	1	1	3	3
	Full 50	1	1	3	2	1	3	2	1	1
	Full 75	3	3	1	1	2	2	4	2	2
	Full 100	4	4	4	4	4	4	3	4	4
MA	Draw 25	1	2	2	2	3	3	2	4	1
	Draw 50	2	1	3	1	2	2	3	2	3
	Draw 75	3	3	1	3	1	1	1	3	2
	Draw 100	4	4	4	4	4	4	4	1	4
	Full 25	1	1	1	1	1	1	1	2	1
	Full 50	3	2	2	4	2	2	4	3	2
	Full 75	2	3	3	2	3	3	2	4	3
	Full 100	4	4	4	3	4	4	3	1	4

between the modeling bias of LASSO and decision trees has significant impact. LASSO is a linear technique and has a regularization pressure parameter making it an interesting option.

Our second investigation explored choosing different sets of subprograms for modeling. Rather than use all the subprograms of one program, it mixed subprograms across a subset of programs from the entire population. Our question was whether identifying useful subprograms in this way and integrating them into selection (via integration of model fitness for a program) and/or genetic variation (via archive based crossover) would yield superior error for the best of run program. Again, we found our results to be equivocal. None of the configurations emerged consistently superior. Again, however ranking and error differed among benchmarks.

This work brings to light particular paths that extend the concepts and understanding of BGP. There are several avenues that could be explored:

1. It would be interesting to see if using a machine learning model whose purpose is more in line with what BGP asks for would benefit the evolutionary process.

For example, instead of building an entire machine learning model on the trace, one could use a feature selection technique, or measure the statistical correlation between the columns. The output would provide material with which to populate the archive. However, this would not provide additional fitness measures.

2. It is unclear exactly why the BGP model that uses the combined traces of all of the programs in the population performed less well than running the model on each program trace independently. It is possible that the idea has merit, but the particulars were not a good fit for BGP. In particular, in each generation only a single machine learning model is built. Therefore, all of the selected trees put into the archive in a single generation have the same weight.

3. In BGP if two subtrees have identical columns in the program trace (i.e. identical *semantics*), only the smaller subtree is kept. This introduces a bias that is not necessarily beneficial to the evolutionary process. It would be interesting to explore how common subtrees with identical semantics are, and if choosing the smaller tree is the better choice.

References

1. David Basin, Yves Deville, Pierre Flener, Andreas Hamfelt, and Jürgen Fischer Nilsson. Synthesis of programs in computational logic. In *PROGRAM DEVELOPMENT IN COMPU-TATIONAL LOGIC*, pages 30–65. Springer, 2004.
2. Robyn Ffrancon and Marc Schoenauer. Memetic semantic genetic programming. In *Proceedings of the 2015 Annual Conference on Genetic and Evolutionary Computation*, GECCO '15, pages 1023–1030, New York, NY, USA, 2015. ACM.
3. Alex Graves, Greg Wayne, Malcolm Reynolds, Tim Harley, Ivo Danihelka, Agnieszka Grabska-Barwińska, Sergio Gómez Colmenarejo, Edward Grefenstette, Tiago Ramalho, John Agapiou, et al. Hybrid computing using a neural network with dynamic external memory. *Nature*, 538(7626):471–476, 2016.
4. Sumit Gulwani. Dimensions in program synthesis. In *Proceedings of the 12th international ACM SIGPLAN symposium on Principles and practice of declarative programming*, pages 13–24. ACM, 2010.
5. Mark Hall, Eibe Frank, Geoffrey Holmes, Bernhard Pfahringer, Peter Reutemann, and Ian H. Witten. The weka data mining software: An update. *SIGKDD Explor. Newsl.*, 11(1):10–18, November 2009.
6. Thomas Haynes. On-line adaptation of search via knowledge reuse. *Genetic Programming 1997: Proceedings of the Second Annual Conference*, pages 156–161. Morgan Kaufmann Publishers Inc., 1997.
7. John R Koza. *Genetic programming: on the programming of computers by means of natural selection*, volume 1. MIT press, 1992.
8. Krzysztof Krawiec. *Behavioral Program Synthesis with Genetic Programming*, volume 618 of *Studies in Computational Intelligence*. Springer International Publishing, 2015.
9. Krzysztof Krawiec and Una-May O'Reilly. Behavioral programming: a broader and more detailed take on semantic gp. In *Proceedings of the 2014 Annual Conference on Genetic and Evolutionary Computation*, pages 935–942. ACM, 2014.
10. Krzysztof Krawiec and Jerry Swan. Pattern-guided genetic programming. In *Proceedings of the 15th annual conference on Genetic and evolutionary computation*, pages 949–956. ACM, 2013.

11. Paweł Liskowski and Krzysztof Krawiec. Online discovery of search objectives for test-based problems. *Evolutionary Computation*, 25:375–406, 2016.
12. James McDermott, David R White, Sean Luke, Luca Manzoni, Mauro Castelli, Leonardo Vanneschi, Wojciech Jaskowski, Krzysztof Krawiec, Robin Harper, Kenneth De Jong, et al. Genetic programming needs better benchmarks. In *Proceedings of the 14th annual conference on Genetic and evolutionary computation*, pages 791–798. ACM, 2012.
13. Robert I McKay. Fitness sharing in genetic programming. In *Proceedings of the 2nd Annual Conference on Genetic and Evolutionary Computation*, pages 435–442. Morgan Kaufmann Publishers Inc., 2000.
14. Nicholas Freitag McPhee, Brian Ohs, and Tyler Hutchison. Semantic building blocks in genetic programming. In *European Conference on Genetic Programming*, pages 134–145. Springer, 2008.
15. Martin C Rinard. Example-driven program synthesis for end-user programming: technical perspective. *Communications of the ACM*, 55(8):96–96, 2012.
16. Armando Solar-Lezama. Program synthesis by sketching. PhD Thesis, University of California, Berkeley, 2008.
17. Saurabh Srivastava, Sumit Gulwani, and Jeffrey S Foster. From program verification to program synthesis. In *ACM Sigplan Notices*, volume 45, pages 313–326. ACM, 2010.
18. Leonardo Vanneschi, Mauro Castelli, and Sara Silva. A survey of semantic methods in genetic programming. *Genetic Programming and Evolvable Machines*, 15(2):195–214, 2014.
19. Westley Weimer, Stephanie Forrest, Claire Le Goues, and ThanhVu Nguyen. Automatic program repair with evolutionary computation. *Communications of the ACM*, 53(5):109–116, 2010.

Chapter 2
Schema Analysis in Tree-Based Genetic Programming

Bogdan Burlacu, Michael Affenzeller, Michael Kommenda,
Gabriel Kronberger, and Stephan Winkler

Abstract In this chapter we adopt the concept of *schemata* from schema theory
and use it to analyze population dynamics in genetic programming for symbolic
regression. We define schemata as tree-based wildcard patterns and we empirically
measure their frequencies in the population at each generation. Our methodology
consists of two steps: in the first step we generate schemata based on genealogical
information about crossover parents and their offspring, according to several
possible schema definitions inspired from existing literature. In the second step, we
calculate the matching individuals for each schema using a tree pattern matching
algorithm. We test our approach on different problem instances and algorithmic
flavors and we investigate the effects of different selection mechanisms on the
identified schemata and their frequencies.

2.1 Introduction

2.1.1 Diversity and Evolutionary Dynamics

"Evolutionary dynamics" is an often-encountered expression in genetic programming (GP) research. It refers to changes within the population, such as quality
and size distribution [15], genotype-phenotype maps and neutral networks [4, 9],

B. Burlacu (✉) · M. Affenzeller · M. Kommenda · S. Winkler
Heuristic and Evolutionary Algorithms Laboratory, University of Applied Sciences Upper
Austria, Hagenberg, Austria

Institute for Formal Models and Verification, Johannes Kepler University, Linz, Austria
e-mail: bogdan.burlacu@fh-hagenberg.at; michael.affenzeller@heuristiclab.com;
michael.kommenda@fh-hagenberg.at; stephan.winkler@fh-hagenberg.at

G. Kronberger
Heuristic and Evolutionary Algorithms Laboratory, University of Applied Sciences Upper
Austria, Hagenberg, Austria
e-mail: gabriel.kronberger@fh-hagenberg.at

© Springer International Publishing AG, part of Springer Nature 2018   17
W. Banzhaf et al. (eds.), *Genetic Programming Theory and Practice XV*,
Genetic and Evolutionary Computation, https://doi.org/10.1007/978-3-319-90512-9_2

diversity [5, 6], modularity and building blocks [11, 25], bloat [18], evolvability [2, 22] or emergent phenomena [3].

The dynamics of the population are uniquely influenced by the interplay between selection and recombination operators (crossover, mutation), as well as specific parameterizations and problem instances. As a biologically-inspired process, GP is able to deal with noisy data, multiple local optima, non-smooth objective functions, while also critically depending on genetic diversity in order to evolve the solution candidates towards the given goal. Population diversity at both the genotypic and phenotypic level remains one of the main focus points for GP research.

In this work we analyze population diversity looking at the distribution of solution candidates into subsets that belong to the same schema or structural template. We define such templates as rooted trees containing wildcard node symbols in their structure. Additionally, we describe population convergence via schema frequency curves over the evolutionary run.

2.1.2   Genetic Programming Schemata

The study of schema theorems began with John Holland's work on providing a mathematical justification for the performance of genetic algorithms. The canonical version of a genetic algorithm (Holland [8]) used a fixed-length binary string encoding where each bit took a value from the set $\{0, 1\}$. Holland then defined schemata (or *schemata*) as binary string templates with symbols from the set $\{0, 1, *\}$, where $*$ represents a wildcard symbol that can be matched by either a 0 or a 1.

The fixed length schemata, each equivalent to a hyperplane in the search space, represented suitable theoretical instruments for the analysis of genetic operators and their effects on the distribution of solution candidates along the hyperplanes, in relation with the average fitness of the population. Holland's schema theory states that the number of low-order, low defining-length schemata with above average fitness increases exponentially between successive generations, where:

- A schema's order is given by the number of fixed positions in the binary string
- The defining length is given by the distance between the first and last fixed positions in the binary string
- Schema average fitness is the average fitness if its matching individuals

In this context, low-order, low-defining-length schemata are seen as *building blocks*, structural patterns increasingly sampled by selection and used by the genetic algorithm to assemble better and better solutions.

It was later shown by Poli [13] that Holland's findings are also valid in the context of GP, with some small provisions: "building blocks in GP and variable-size GAs with one-point crossover exist, but they are not necessarily all short, low-order or highly fit". Schema theorems for GP are complicated by the variable-length tree encoding, requiring mathematical formulations for the expected schema frequencies to also account for the size variation of individuals under the action of selection,

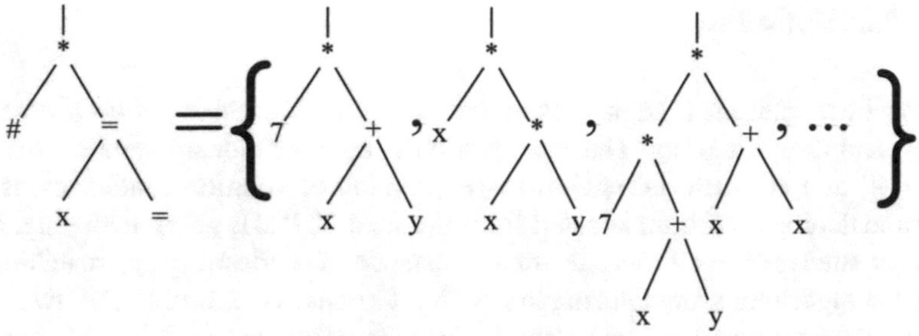

Fig. 2.1 Example hyperschema and matching tree individuals [12]

crossover and mutation. Several schema definitions dealing with these issues were proposed in the literature [24].

Despite significant progress in the last couple of decades, leading to exact formulations of the expected number of individuals sampling a schema at the next generation [16, 17], "large gaps remain between GP theory and practice", due to the large number of schema equations in typical GP populations, and the "large number of terms growing proportionally to the square of the number of program shapes times the square of the number of possible crossover points" [19]. Thus, from a practical perspective, the application of schema theory on concrete algorithms and problem instances remains problematic.

In this work we attempt to close the gap between schemata as theoretical instruments for the analysis of population dynamics and their role in empirical investigations and we introduce a practical methodology to identify GP schemata and compute their frequencies. We consider the hyperschema definition by Poli et al. [12], where a schema is a rooted tree template that may include two types of wildcard symbols:

- The '=' symbol matches any valid node of the same type and arity
- The '#' symbol matches any valid subtree

Figure 2.1 shows an example of a hyperschema and matching trees. We notice that the # symbol can match both leaf and function nodes, while the = symbol only matches nodes of the same type (a function node and a leaf node, respectively, for the two occurrences below).

Originally, the set of wildcards in Poli's hyperschema was chosen in such a way as to make it easier to evaluate the effects of the genetic operators on schema frequencies and to enable a more concise mathematical representation of the schema equations. Our proposed methodology is not bound by such considerations and supports different schema structures, containing either or both wildcard symbols.

The remainder of the chapter is organized as follows: Sect. 2.2 describes our methodology for schema generation and matching, Sect. 2.3 gives details about our empirical experiments, Sect. 2.4 shows the obtained results and Sect. 2.5 discusses some final conclusions.

2.2   Methodology

We construct relevant schemata using hereditary relationships between crossover
parents and their offspring. The schemata may include wildcard symbols from the
set {=, #} and are matched against the population of solution candidates using a
pattern matching algorithm adapted from the field of XML query matching. Since
GP schemata represent a more restricted instance of wildcard query matching, we
adapt the algorithm's implementation with additional constraints. The two steps,
schema generation and schema matching, are described in more detail below. The
methodology was implemented in HeuristicLab [23].

2.2.1   Schema Generation

Conceptually, we expand on the idea by Stephens and Waelbroek [20] that "at the
level of the microscopic degrees of freedom, the strings, the action of crossover by
its very nature introduces the notion of a schema."

The schema generation algorithm tries to exploit the fact that structural similarity
is passed on (to various degrees) from parents to their offspring via the crossover
operation. Additionally, it is assumed that successful individuals selected for
reproduction will participate as root parents[1] in multiple crossover operations. In
these circumstances, we can generate schemata from crossover root parents by
considering crossover cutpoints as potential candidates for wildcard placement. We
arrive at the following heuristic:

1. Group individuals based on their common root parent
2. Identify all genetic fragments and their respective positions in the root parent
3. For each fragment f with preorder index f_i in the root parent, replace the node
   at position f_i with a wildcard.

The heuristic is controlled by a minimum schema length parameter which limits
wildcard placement in order to avoid the creation of 'match-all' schemata (schemata
that contain wildcards in the tree root or in its close proximity). The method is listed
as pseudocode in Algorithm 2.1.

Since wildcards are inserted at cutpoint locations, the structure of the generated
schemata is influenced indirectly by the selection pressure applied on the population,
which determines the multiplicity of root parent individuals (how many times each
individual participates in crossover as a root parent) and therefore the number of
wildcards. Intuitively, the method will generate more general schemata under high

---

[1]The terms *root parent* and *non-root parent* refer to the two parents involved in a crossover
operation: the root parent passes on to the child its entire rooted tree structure, with the exception
of the subtree swapped by crossover at an arbitrary location (called a cutpoint) from the non-root
parent.

selection pressure and more specific schemata (containing fewer wildcards) under lower selection pressure. The algorithm can generate different kinds of schemata (according to the schema definitions in the literature, for an excellent summary see [24]), depending on the kind of wildcard symbols used for replacement.

Algorithm 2.1 Schema generation

Method GenerateSchemas (*genealogy graph, minimum schema length*)

schemas ← new list;   // list holding the generated schemas
// use genealogy information to group offspring with common
  parents
1   group all children of the current generation based on their common root parent **foreach** *root parent p* **do**
2   **if** *length(p) < minimum schema length* **then**
3 | **continue**
4   schema ← copy of *p* replaced ← false indexes ← preorder indices of the crossover cutpoints in all children sort(indexes);   // sort indices by cutpoint level in descending order
5   **foreach** *index i from* indexes **do**
6   subtree ← the subtree at position i in schema   **if** *length(*schema*) − length(*subtree*) +1 < minimum schema length* **then**
7 | **continue**
8   replacement ← new wildcard node;   // either = or #
9   ReplaceSubtree(subtree, replacement);   // replace the subtree with the wildcard in the parent's structure
10 | replaced ← true
11   **if** *replaced* **then** if the schema contains at least one wildcard
12 | add schema to schemas
13   **return** schemas

2.2.2 Schema Matching

The schema matching part of our methodology is based on the algorithm for the tree homeomorphism decision problem by Götz et al. [7], which tries to find a non-injective mapping between every parent-child pair in a query tree Q (the schema) and corresponding ancestor-descendant pairs in data tree D (the matched individual). Such a situation is shown in Fig. 2.2 where, according to the algorithm, the query tree Q is matched by the data tree D. The algorithm runs in $O(|D| \cdot |Q| \cdot depth(Q))$ time using a stack of depth bounded by $O(depth(D) \cdot branch(D))$.

We notice that the algorithm in its default implementation does not enforce strict enough matching rules as required by schema matching, since tree nodes are matched from the bottom up if they have the same label, without additional considerations for their depth in the tree (relative to the root node). Therefore we

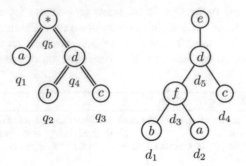

Fig. 2.2 Example query matching between query tree Q (left) and data tree D (right) [7]. The algorithm finds a non-injective mapping between every parent-child pair in Q and a corresponding ancestor-descendant pair in D. The answer will be *yes* (there is a matching) if, starting from the bottom up, the procedure can map the root nodes of the two trees (q_5 and d_6)

added additional rules in our implementation, to make sure two nodes are only matched if they are on the same level in the tree and their parent and children nodes are matched as well. Another important detail is the matching of commutative symbols, in which case the algorithm does not consider the order of the child subtrees (internally, a sorting is performed). For example, a schema $(+ = x)$ (in postfix notation) will be matched by an individual $(+ x y)$ because the $+$ symbol is commutative, despite the fact that the x symbol is found at different positions in the argument order.

2.3 Experimental Setup

We compared the evolution of schema frequencies between two algorithmic variants: standard genetic programming (SGP) [10] and genetic programming with strict offspring selection (OSGP) [1]. The difference between the two algorithms consists of an extra selection step enforced by OSGP on the generated offspring, such that offspring get rejected if they do not fulfil certain performance criteria. In effect, the extra selection step concentrates the algorithm's efforts on generating adaptive changes (that do not decrease fitness), making it possible for less fit individuals to participate as parents if they can produce children fitter than themselves, while high fitness individuals might not contribute if they cannot be improved.

Each problem and algorithm configuration was repeated for a number of 20 runs, from which a single representative run was selected based on best performance on the training data. This final run selection step was necessary for clarity and space reasons, as the slight differences between runs (particularly at the genotypic level) make it impossible for schemata generated from one population genealogy to be applied to another population genealogy.

Table 2.1 SGP configuration

Population size	500 individuals (SGP), 200 individuals (OSGP)
Termination criteria	SGP: 100 generations
	OSGP: maximum selection pressure 100
Selection mechanism	SGP: Tournament (group size 5) and proportional
	OSGP: Gender specific (proportional + random)
Crossover probability	100%, subtree crossover
Mutation probability	25%, evenly divided between single-, multi-point,
	or change node type mutation
Function set	$F = \{\times, \div, +, -\}$
Terminal set	$T = \{\text{weighted variables, ephemeral constants}\}$
Maximum tree depth	12
Maximum tree length	25

Table 2.2 OSGP configuration

Population size	200 individuals
Max selection pressure	100 (acting as termination criteria)
Parent selection	Gender specific (proportional + random)
Offspring selection	Strict (offspring must outperform parents)

2.3.1 Algorithm Parameters

We applied our schema generation and matching methodology at each generation on the whole population of solution candidates.[2] The parameterizations for the two algorithms are presented in Tables 2.1 and 2.2.

The OSGP algorithm uses the same primitive set, tree depth and size limits and crossover and mutation operators as SGP, with differences in population size, stopping criteria and selection mechanism.

2.3.2 Problem Instances

For this experiment, we selected one symbolic regression benchmark problem that facilitates discernible genotypic representations of solutions, in order to more easily observe solution fragments or building blocks contained by the schemata. We used the *Poly-10* [14] synthetic symbolic regression benchmark, where the goal is to find the target function:

$$f(\mathbf{x}) = x_1x_2 + x_3x_4 + x_5x_6 + x_1x_7x_9 + x_3x_6x_{10} \tag{2.1}$$

---

[2]To maintain low computational times, certain compromises had to be made in terms of population size and number of generations.

For the second test problem we used the *Tower* dataset [21], containing real-world data in the form of gas chromatography measurements of the composition of a distillation tower.

2.3.3 Analysis Methods

We perform our analysis *a posteriori* with the help of a complete genealogical record of the algorithmic run. We generate a set of potential schemata from the population at each generation, and match it against the whole genealogy in order to determine the evolution of schema frequencies over time.

As diversity loss in the course of the evolutionary process reflects itself in the set of schemata obtained each generation (which can contain duplicates or can repeat structures obtained in previous generations), we additionally perform filtering based on the schema frequency curves. If two schemata have highly correlated frequency curves (with a Pearson's R^2 correlation coefficient value >0.99), one of them is removed from the set of all schemata.

2.4 Empirical Results

We prefix each tested configuration with the name of the algorithm, followed by distinctive parameters such as the selection mechanism and maximum tree length, and then followed by the problem name. For example, the name SGP-P-25-Poly10 denotes a standard GP run with proportional selection and a maximum tree length of 25, while SGP-T-25-Poly10 denodes the same configuration with tournament selection instead.

When discussing schema frequencies, we use the notation $S_{1,P} \ldots S_{10,P}$ for schemata generated by SGP with proportional selection, and $S_{1,T} \ldots S_{10,T}$ for SGP with tournament selection. For OSGP, we use the notation $S_{1,G} \ldots S_{10,G}$ to denote the 10 most common schemata. To keep a concise notation, we repeat the same notation in each section corresponding to each tested problem.

2.4.1 Standard GP

2.4.1.1 Poly-10 Problem

We first look at the convergence of SGP-P-25-Poly10. At the structural level, convergence should manifest itself as an increased occurrence count of repeated patterns in the population. Table 2.3 shows the most frequent schemata found in the

Table 2.3 SGP-P-25-Poly10: most common schemata in the last generation

#	Prefix representation	N (%)
$S_{1,P}$	$(= (+ (* X_3\ X_4)\ (= X_3\ X_4))\ (= (= X_3\ X_4)\ (* X_3\ X_4)))$	0.34
$S_{2,P}$	$(= \#\ (+ (* X_3\ X_4)\ (* X_3\ X_4)))$	0.22
$S_{3,P}$	$(= (= X_2\ \#)\ (* X_2\ X_1))$	0.19
$S_{4,P}$	$(+ (* X_2\ X_1)\ (= X_2\ X_1))$	0.18
$S_{5,P}$	$(+ (= X_3\ X_4)\ (= \#\ X_4))$	0.18
$S_{6,P}$	$(+ (+ (* X_3\ X_4)\ (* X_3\ X_4))\ (+ (* X_3\ X_4)\ (= X_3\ X_4)))$	0.15
$S_{7,P}$	$(+ (+ (* X_6\ X_5)\ (* X_6\ X_5))\ (= (* X_6\ X_5)\ (* X_6\ X_5)))$	0.15
$S_{8,P}$	$(= (* X_6\ X_5)\ (= X_6\ X_5))$	0.15
$S_{9,P}$	$(= \#\ (+ (* X_2\ X_1)\ (* X_2\ X_1)))$	0.14
$S_{10,P}$	$(+ (+ (* X_6\ X_5)\ (= X_6\ X_5))\ (+ (* X_6\ X_5)\ (= X_6\ X_5)))$	0.14

last generation, represented in postfix notation. The notation $S_{1,P} \ldots S_{10,P}$ in the first column of the table is used to designate the schemata obtained in the SGP run with proportional selection.

We notice that some schemata (for example, $S_{1,P}$ and $S_{2,P}$, as well as $S_{3,P}$ and $S_{4,P}$) share a degree of structural similarity. A closer look at their respective frequency curves (not detailed here for space reasons) reveals that:

- The frequency curves for $S_{1,P}$ and $S_{2,P}$ are highly correlated ($R^2 = 0.962$), however $S_{2,P}$ represents a more specific template which matches fewer individuals at each generation.
- The frequency curves for $S_{3,P}$ and $S_{4,P}$ are correlated ($R^2 = 0.916$). In this case $S_{4,P}$ represents the slightly more specific template, matching fewer individuals than $S_{3,P}$.
- The frequency curves for $S_{7,P}$ and $S_{10,P}$ are correlated ($R^2 = 0.907$), with S_{10} being the slightly more specific schema.

The fact that we obtained similar and frequency-correlated schemata via our crossover-based generation procedure indicates the presence of similar parent individuals in the population, suggesting loss of diversity. We focus on the most relevant schemata ($S_{1,P}$, $S_{3,P}$ and $S_{7,P}$) and show their frequency evolution in Fig. 2.3.

The frequency curves show the moments when the algorithm was able to discover parts of the formula such as x_1x_2, x_3x_4 and x_5x_6. The schemata sharply increase their frequency in the population in the beginning of the run and then vary according to the internal dynamics of the evolutionary search (competition between schemata, stagnation in the later stages).

From a diversity perspective, the schema frequency approach has the ability to identify high level similarities in the population (e.g., when 30% of the population share the same genetic template) that would otherwise be hard to notice with conventional metrics like tree distances.

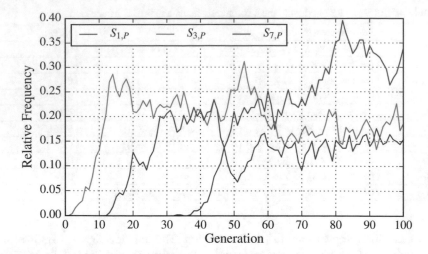

Fig. 2.3 SGP-P-25-Poly10: frequency evolution of relevant schemata

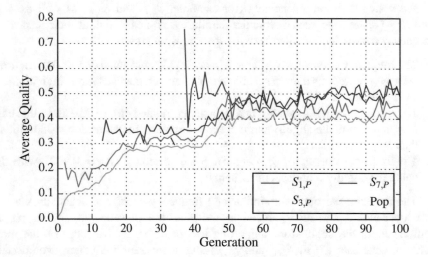

Fig. 2.4 SGP-P-25-Poly10: average population and schema qualities

The results so far confirm that schemata identified by our method correspond to what could be considered as building blocks for this problem, including in their structure the terms of the formula and showing an exponential increase in frequency from the moment of their occurrence.

We calculate schema average quality (as the average quality of their matching individuals) and show the results in Fig. 2.4. The quality curves suggest that the identified schemata are of above-average quality.

Table 2.4 SGP-T-25-Poly10: most common schemata in the last generation

#	Prefix representation	N (%)
$S_{1,T}$	$(= (= (* (* X3 C) X4) (* X5 X6))(= (* (* X3 C) X4) (= (* X3 C) X4)))$	0.38
$S_{2,T}$	$(- (+ (* (* X3 C) X4)(* X5 \#)) (* \# (* X2 X1)))$	0.27
$S_{3,T}$	$(- (+ (* (= X3 C) X4) (* X5 X6)) (* (* X2 X1) (= \# X1)))$	0.25
$S_{4,T}$	$(- (= (= (* X3 C) X4) (= X5 X6)) (* (- (* \# \#) C) (* X2 X1)))$	0.17
$S_{5,T}$	$(- (+ (* X5 X6) (* X5 X6)) (* (- (* (* X2 X1)(* \# X1)) C) (* X2 X1)))$	0.08
$S_{6,T}$	$(- (+ (* (* X3 C) X4) (= X5 \#)) (* (- (* (= X2 X1)(* X2 X1)) C)$ $(= X2 X1)))$	0.07
$S_{7,T}$	$(- (+ (= (* X3 C) X4) (* X5 X6)) (= (= (* (= X2 X1)(* X2 X1)) C)$ $(* X2 X1)))$	0.06
$S_{8,T}$	$(- (= (= (* X3 C) X4) (* X5 \#)) (= (- (* (* X2 X1) (* X2 X1)) \#)$ $(= X2 X1)))$	0.06
$S_{9,T}$	$(- (+ (* (* X3 C) X4) (* X5 X6)) (* (= X2 X1) (- (* (* X3 \#) \#) C)))$	0.06
$S_{10,T}$	$(- (= (* (* X3 C) X4) (* X5 X6)) (= (- (* (* X2 X1) (* X2 X1)) C)$ $(* X2 X1)))$	0.06

A similar situation can be observed for SGP with tournament selection, where several frequent schemata are present in the last generation and shown in Table 2.4. We notice that the top four schemata are matching relatively high proportions of the population and show their detailed frequency evolution in Fig. 2.5.

The generated schemata correspond to solution building blocks, containing terms of the target formula. Compared to proportional selection, the extra selection pressure applied on the population by the tournament selection (with a group size of 5) leads to larger schemata.

The observed schema frequency evolutions for SGP with proportional and tournament selection support the idea that relevant schemata increase in frequency over the generations.

Quality measurements in Fig. 2.6 show a significant difference between the average quality of the population and the average schema qualities. The discontinued line segments in this figure correspond to generations when the schema frequency dropped to zero, therefore an average quality could not be calculated. The results suggest that tournament selection (applying higher pressure on the population) promotes higher quality schemata.

2.4.1.2 Tower Problem

We compare the two standard GP configurations using proportional and tournament selection, denoted SGP-P-25-Tower and SGP-T-25-Tower. The most common schemata for the SGP variant with proportional selection are given in Table 2.5.

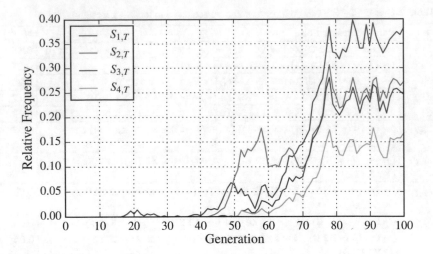

Fig. 2.5 SGP-T-25-Poly10: frequency evolution of relevant schemata

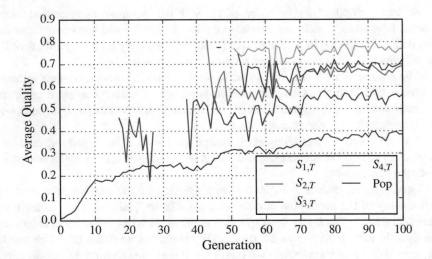

Fig. 2.6 SGP-T-25-Poly10: average population and schema qualities

The obtained templates have low length and only include 2 out of 25 input variables, with the most common schema matching 15% of the population in the last generation. This result suggests that the two variables x_1 and x_6 are more relevant (in terms of the implicit variable ranking performed by GP) for the modeling of the target. In terms of quality, the produced symbolic regression solution achieved a Pearson's R^2 correlation with the target variable of 0.8. As with the previous problem, we plot the evolution of schema frequencies, using a correlation-based

Table 2.5 SGP-P-25-Tower: most common schemata in the last generation

#	Prefix representation	N (%)
$S_{1,P}$	(= (= X6 X1) (= # X1))	0.15
$S_{2,P}$	(= (= X6 C) (= X6 C))	0.14
$S_{3,P}$	(= (= C #) (= C X6))	0.12
$S_{4,P}$	(= (= X6 X1) (− X6 X1))	0.12
$S_{5,P}$	(= (− X6 X1) (= X6 X1))	0.11
$S_{6,P}$	(= (= X6 C) (= C X6))	0.11
$S_{7,P}$	(= (= X6 X6) (= X6 X6))	0.09
$S_{8,P}$	(= (∗ C X6) (= C X6))	0.07
$S_{9,P}$	(= (= X6 C) (∗ C X6))	0.07
$S_{10,P}$	(= X1 (= (= X6 #) X6))	0.06

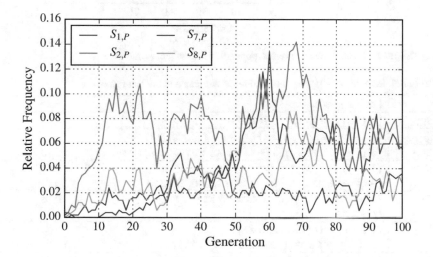

Fig. 2.7 SGP-P-25-Tower: frequency evolution of relevant schemata

filtering step to eliminate similar curves. The Pearson's R^2 correlation values for $S_{1,P} \ldots S_{10,P}$ show that:

- $S_{1,P}$ is highly correlated with $S_{4,P}$ and $S_{5,P}$
- $S_{2,P}$ is highly correlated with $S_{6,P}$
- $S_{8,P}$ is highly correlated with $S_{9,P}$

The frequencies of the remaining schemata are shown in Fig. 2.7, while their qualities, along with the average quality of the population are shown in Fig. 2.8. We see that $S_{2,P}$ becomes frequent rather early and is overall more frequent that $S_{1,P}$, while the latter has a marginally higher frequency in the last generation. Quality-wise, $S_{1,P}$ and $S_{2,P}$ are clearly above the average of the population, while $S_{7,P}$ and $S_{8,P}$ occasionally dip below the average.

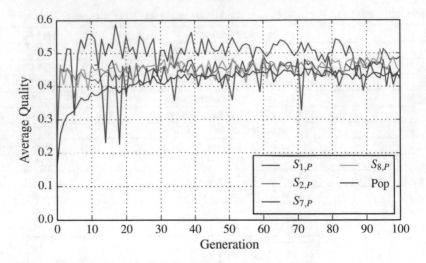

Fig. 2.8 SGP-P-25-Tower: average population and schema qualities

Table 2.6 SGP-T-25-Tower: most common schemata in the last generation

#	Prefix representation	Freq.
$S_{1,T}$	$(+ (- (/ (* \# \text{X23}) \text{X1}) \#) \text{X12})$	0.33
$S_{2,T}$	$(+ (- (= (* \text{X4 X23}) \text{X1}) (= (= \# \text{X23}) \text{X1})) \text{X12})$	0.28
$S_{3,T}$	$(+ (- (/ (* \# \text{X23}) \text{X1}) (- (/ (= \text{X6 X23}) \text{X1}) \#)) \text{X12})$	0.20
$S_{4,T}$	$(+ (- (/ (* \text{X4 X23}) \text{X1})$ $(- (/ (* \text{X6 X23}) \text{X1}) (= (* \text{X6 X23}) \text{X1}))) \text{X12})$	0.17
$S_{5,T}$	$(+ (= (= \# \text{X1}) (- (/ (* \text{X6 X23}) \text{X1})$ $(/ (* \# \text{X23}) \text{X1}))) \text{X12})$	0.12
$S_{6,T}$	$(+ (- \# (= (/ (* \text{X6 X23}) \text{X1})$ $((- (= (* \text{X6 X23}) (* \text{X6 X23})) \text{X1}))) \text{X12})$	0.08
$S_{7,T}$	$(+ (- (= (* \# \text{X23}) \text{X1}) (= (/ (* \text{X6 X23}) \text{X1})$ $(- (/ (* \text{X6 X23}) (* \# \text{X23})) \text{X1}))) \text{X12})$	0.07
$S_{8,T}$	$(+ (- (/ (= \text{X4 } \#) \text{X1}) (- (= (= \text{X6 } \#) \text{X1})$ $(= (/ (= \text{X6 X23}) (* \text{X6 X23})) \text{X1}))) \text{X12})$	0.07
$S_{9,T}$	$(= (/ (* \text{X6 X23}) \text{X1}) (- (/ (* \# \text{X23}) (= \text{X6 } \#)) \text{X1}))) \text{X12})$ $(+ (- (= (* \text{X4 X23}) \text{X1})$	0.07
$S_{10,T}$	$(+ (= (/ (* \text{X4 X23}) \text{X1})$ $(= (/ (* \text{X6 X23}) \text{X1}) (- (= (= \text{X6 X23}) (= \# \#)) \#))) \text{X12})$	0.07

Tournament selection determines the evolution of more complex schemata. The ten most frequent schemata in the last generation shown in Table 2.6 are larger in size, match more individuals and contain more variables from the dataset.

Correlation analysis of the frequency curves reveals that:

- $S_{1,T}$ is highly correlated with $S_{2,T}$, $S_{3,T}$ and $S_{4,T}$ with an R^2 value of 0.96.
- $S_{7,T}$ is highly correlated with $S_{10,T}$ with an R^2 value of 0.95

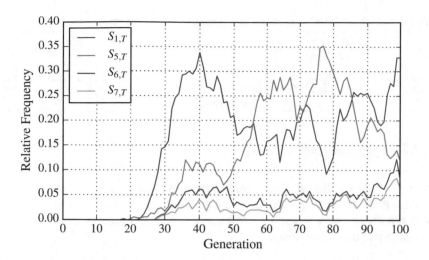

Fig. 2.9 SGP-T-25-Tower: frequency evolution of relevant schemata

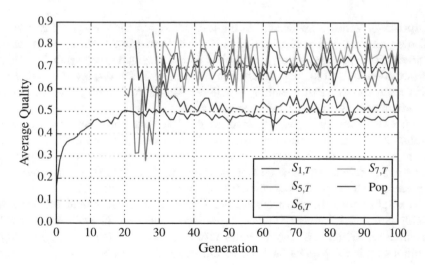

Fig. 2.10 SGP-T-25-Tower: average population and schema qualities

Filtering correlated schemata, we display the remaining schema frequency curves in Fig. 2.9. Interestingly $S_{1,T}$, the most common schema in the last generation has a noticeably lower average quality compared to $S_{5,T}$, $S_{6,T}$ and $S_{7,T}$, although it still manages to rise above the average population quality, as seen in Fig. 2.10.

Table 2.7 OSGP-25-Poly10: most common schemata in the last generation

#	Prefix representation	N (%)
$S_{1,G}$	$(- (- (* X3\ X4)\ (* (+ X7\ \#)$ $(* X6\ (+ (+ (* X3\ X10)\ X5)\ X5))))\ (= X1\ X2))$	1
$S_{2,G}$	$(- (- (* X3\ X4)\ (= (+ X7\ (+ X1\ C))$ $(* X6\ (= (= (* X3\ X10)\ X5)\ X5))))\ (* X1\ X2))$	1

2.4.2 Offspring Selection GP

As previously mentioned, OSGP implements an additional selection step which decides if the offspring produced by mutation and crossover are accepted into the next generation. We analyze the influence of offspring selection on the generated schemata and their frequencies.

2.4.2.1 Poly-10 Problem

Surprisingly, schema frequencies in the last generation show that only two out of all the generated schemata managed to survive. Furthermore, these two schemata represent a very similar genotypic template which managed to propagate itself to all of the individuals in the population. The two schemata are displayed in Table 2.7. This result shows that it is entirely possible under strict offspring selection for the algorithm to converge to a single genetic template.

Since we only have two schemata in the last generation, we investigate the evolution of schema frequencies using a different strategy: we rank the schemata based on their overall frequency, that is, the average of their individual frequencies in each generation. The new ranking is shown in Table 2.8, where the frequency represents an average of the schema frequency over all generations.

Several of the schemata from Table 2.8 match the same individuals and have highly correlated frequency curves. These schemata were filtered from Fig. 2.11 to eliminate clutter. The figure shows multiple schemata ($S_{3,G}$, $S_{6,G}$ and $S_{8,G}$) proliferating in the population in the earlier generations of the run, only to become extinct later.

After generation 38, the two most frequent schemata in the last generation, $S_{1,G}$ and $S_{2,G}$ have overlapping frequency curves, suggesting that $S_{2,G}$ has a higher degree of specificity, presumably due to the lack of '#' wildcard symbols in its structure.

2.4.2.2 Tower Problem

We notice a similar behavior for the *Tower* problem, where a single schema denoted as $S_{1,G}$ matches all the individuals in the last generation:

$(/ C (- X1 (* C (- X12 (/ (- (/ (+ (* X6\ X23) (= X22\ \#)) C) X5) (- X1\ C))))))$

Like before, we consider in this situation the most frequent schemata overall, shown in Table 2.9.

Table 2.8 `OSGP-25-Poly10`: most common schemata overall

#	Prefix representation	O (%)
$S_{1,G}$	(− (− (∗ X3 X4) (∗ (+ X7 #) (∗ X6 (+ (+ (∗ X3 X10) X5) X5)))) (= X1 X2))	0.46
$S_{2,G}$	(− (− (∗ X3 X4) (= (+ X7 (+ X1 C)) (∗ X6 (= (= (∗ X3 X10) X5) X5)))) (∗ X1 X2))	0.45
$S_{3,G}$	(− (− (∗ X3 X4) (∗ (+ # C) (∗ X6 (+ (+ (∗ X3 X10) X5) X5)))) (∗ X1 X2))	0.22
$S_{4,G}$	(− (− (∗ X3 X4) (∗ (+ (= X1 #) C) (∗ X6 (+ (+ (∗ X3 X10) X5) X5)))) (∗ X1 X2))	0.15
$S_{5,G}$	(− (− (= X3 X4) (∗ (+ (+ X1 X7) C) (∗ X6 (= (= (= X3 X10) X5) X5)))) (∗ X1 X2))	0.14
$S_{6,G}$	(− (− (∗ X3 X4) (∗ (+ C (= # #)) (∗ X6 (+ (∗ X1 X2) X5)))) (∗ X1 X2))	0.10
$S_{7,G}$	(− (− (= X3 X4) (∗ (+ X1 C) (∗ X6 (+ (+ (∗ X3 X10) X5) X5)))) (= X1 X2))	0.05
$S_{8,G}$	(− (− (= X3 X4) (∗ (+ # C) (∗ X6 X5))) (∗ X1 X2))	0.05
$S_{9,G}$	(− (− (∗ X3 X4) (∗ (+ C (∗ X5 (∗ X2 X2))) (∗ X6 (+ (= X1 X2) X5)))) (= # X2))	0.04
$S_{10,G}$	(− (− (= X3 X4) (∗ (+ C (= X1 #)) (∗ X6 (+ (∗ X1 X2) X5)))) (= X1 X2))	0.03

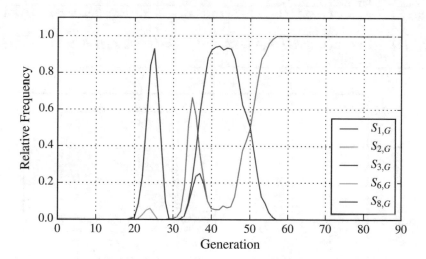

Fig. 2.11 `OSGP-25-Poly10`: frequency evolution of relevant schemata

Figure 2.12 shows the evolution of schema frequencies for the top three most frequent schemata from Table 2.9. We see schema $S_{1,G}$ rising in frequency after generation 20 and driving other schemata to extinction.

Compared to SGP, the schemata obtained by OSGP and their frequency evolution suggests a more pronounced loss of diversity as the population becomes dominated by a single schema.

Table 2.9 OSGP-25-Tower: most common schemata overall

#	Prefix representation	Freq.
$S_{1,G}$	(/ C (− X1 (∗ C (− X12 (/ (− (/ (+ (∗ X6 X23) (= X22 #)) C) X5) (− X1 C))))))	0.70
$S_{2,G}$	(− X1 (∗ C (− X12 (/ (− (/ (+ (∗ X6 X23) (− (= X22 #) X2)) C) X5) (− X1 C)))))	0.03
$S_{3,G}$	(− X1 (∗ C (− X12 (/ (/ (+ (∗ X6 X23) (− (− (− X22 #) #) X2)) C) (= X1 C)))))	0.02
$S_{4,G}$	(− X1 (∗ C (− X12 (/ (/ (+ (∗ X6 X23) (− (− (= X22 #) X2) X2)) C) (− X1 C)))))	0.01
$S_{5,G}$	(− X1 (∗ C (− X12 (/ (/ (+ (∗ X6 X23) (− (= (− X22 C) #) X2)) C) (− X1 C)))))	0.01
$S_{6,G}$	(− X1 (∗ C (− X12 (/ (− (/ (+ (= X6 X23) (− (= X22 C) X2)) C) X5) (− X1 C)))))	0.007
$S_{7,G}$	(− X1 (∗ C (− X12 (/ (/ (= (+ (∗ X6 X23) (− X22 #)) C) C) (− X1 C)))))	0.007
$S_{8,G}$	(− X1 (∗ C (− X12 (/ (/ (+ (= X6 X23) (− X22 #)) C) (− X1 C)))))	0.006
$S_{9,G}$	(− X1 (∗ C (− X12 (/ (− (/ (+ (∗ X6 X23) (− (− X22 X2) X2)) C) X5) (= X1 C)))))	0.006
$S_{10,G}$	(− X1 (∗ C (= X12 (/ (/ (+ (∗ X6 X23) (− (= (− X22 X23) #) X2)) C) (− X1 C)))))	0.006

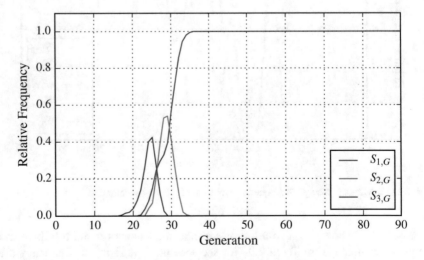

Fig. 2.12 OSGP-25-Tower: frequency evolution of relevant schemata

2.5 Conclusion

We described in this chapter a practical approach for performing schema analysis on GP populations, considering a well-known schema definition (Poli's hyperschema) that uses two types of wildcard symbols for function and leaf nodes, respectively. The methodology can be easily extended to include different schema definitions or stricter matching rules.

Hyperschema are generated algorithmically by taking into account genealogical information about crossover offspring and their respective parents. A pattern matching algorithm is then used to match schemata against the GP population at each generation.

We tested our methodology using two test problems (*Poly-10* and *Tower*) and two algorithmic variants: Standard GP and Offspring Selection GP. The results validate our approach: the identified schemata for each test problem are of increasing frequency in the population and above-average quality. Compared to other methods for measuring genotypic diversity, our schema-based approach offers a detailed picture of the propagation of repeated patterns, while also being able to identify these patterns.

The evolution of schema frequencies suggests that diversity loss starts to occur early in the evolutionary run and tends to homogenize the genotypic structure of the population. As expected, this phenomenon is highly influenced by the selection mechanism. For both problems, the SGP runs using tournament selection displayed lengthier, more frequent and more specific schemata. Offspring selection determines even more drastic effects, as the population shares a single (and rather specific) genetic template.

Future research in this direction will focus on a more detailed analysis of population dynamics where we also consider schema disruption events. The approach can also be employed online to guide the evolutionary process, for example by avoiding loss of diversity via localized mutation rates within frequent schemata.

Acknowledgements The work described in this paper was done within the COMET Project Heuristic Optimization in Production and Logistics (HOPL), #843532 funded by the Austrian Research Promotion Agency (FFG).

References

1. Affenzeller, M., Winkler, S., Wagner, S., Beham, A.: Genetic Algorithms and Genetic Programming: Modern Concepts and Practical Applications. Numerical Insights. CRC Press, Singapore (2009)
2. Altenberg, L., et al.: The evolution of evolvability in genetic programming. Advances in genetic programming **3**, 47–74 (1994)
3. Banzhaf, W.: Genetic programming and emergence. Genetic Programming and Evolvable Machines **15**(1), 63–73 (2014). https://doi.org/10.1007/s10710-013-9196-7

4. Banzhaf, W., Leier, A.: Evolution on neutral networks in genetic programming. In: Genetic programming theory and practice III, pp. 207–221. Springer (2006)
5. Burke, E., Gustafson, S., Kendall, G.: A survey and analysis of diversity measures in genetic programming. In: Proceedings of the 4th Annual Conference on Genetic and Evolutionary Computation, pp. 716–723. Morgan Kaufmann Publishers Inc. (2002)
6. Burke, E.K., Gustafson, S., Kendall, G.: Diversity in genetic programming: An analysis of measures and correlation with fitness. IEEE Transactions on Evolutionary Computation **8**(1), 47–62 (2004)
7. Götz, M., Koch, C., Martens, W.: Efficient algorithms for descendant-only tree pattern queries. Inf. Syst. **34**(7), 602–623 (2009). https://doi.org/10.1016/j.is.2009.03.010
8. Holland, J.H.: Adaptation in Natural and Artificial Systems. The University of Michigan Press (1975)
9. Hu, T., Banzhaf, W., Moore, J.H.: Population Exploration on Genotype Networks in Genetic Programming. In: Proceedings of the 13th International Conference on Parallel Problem Solving from Nature – PPSN XIII, 2014, pp. 424–433. Springer International Publishing, Cham (2014)
10. Koza, J.R.: Genetic Programming: On the Programming of Computers by Means of Natural Selection. MIT Press, Cambridge, MA, USA (1992)
11. Krawiec, K., Wieloch, B.: Functional modularity for genetic programming. In: Proceedings of the 11th Annual Conference on Genetic and Evolutionary Computation, GECCO '09, pp. 995–1002. ACM, New York, NY, USA (2009). http://doi.acm.org/10.1145/1569901.1570037
12. Poli, R.: Hyperschema theory for gp with one-point crossover, building blocks, and some new results in ga theory. In: Genetic Programming, Proceedings of EuroGP 2000, pp. 15–16. Springer-Verlag (2000)
13. Poli, R.: Exact schema theory for genetic programming and variable-length genetic algorithms with one-point crossover. Genetic Programming and Evolvable Machines **2**(2), 123–163 (2001). https://doi.org/10.1023/A:1011552313821
14. Poli, R.: A simple but theoretically-motivated method to control bloat in genetic programming. In: Proceedings of the 6th European Conference on Genetic Programming, EuroGP'03, pp. 204–217. Springer-Verlag, Berlin, Heidelberg (2003). http://dl.acm.org/citation.cfm?id=1762668.1762688
15. Poli, R., Langdon, W.B., Dignum, S.: Generalisation of the limiting distribution of program sizes in tree-based genetic programming and analysis of its effects on bloat. In: in GECCO 2007: Proceedings of the 9th Annual Conference on Genetic and Evolutionary, pp. 1588–1595. ACM Press (2007)
16. Poli, R., McPhee, N.F.: General schema theory for genetic programming with subtree-swapping crossover: Part I. Evolutionary Computation **11**(1), 53–66 (2003).
17. Poli, R., McPhee, N.F.: General schema theory for genetic programming with subtree-swapping crossover: Part II. Evolutionary Computation **11**(2), 169–206 (2003). https://doi.org/10.1162/106365603766646825
18. Poli, R., McPhee, N.F.: Covariant parsimony pressure for genetic programming. In: GECCO 2008: Proceedings of the 10th annual conference on Genetic and Evolutionary Computation, pp. 1267–1274. ACM Press (2008)
19. Poli, R., Vanneschi, L., Langdon, W.B., McPhee, N.F.: Theoretical results in genetic programming: The next ten years? Genetic Programming and Evolvable Machines **11**(3–4), 285–320 (2010). http://dx.doi.org/10.1007/s10710-010-9110-5
20. Stephens, C.R., Waelbroeck, H.: Effective degrees of freedom in genetic algorithms. Physical Review E **57**(3), 3251–3264 (1998)
21. Vladislavleva, E.J., Smits, G.F., Den Hertog, D.: Order of nonlinearity as a complexity measure for models generated by symbolic regression via pareto genetic programming. Evolutionary Computation, IEEE Transactions on **13**(2), 333–349 (2009)
22. Wagner, G.P., Altenberg, L.: Perspective: complex adaptations and the evolution of evolvability. Evolution **50**, 967–976 (1996)

23. Wagner, S., Kronberger, G., Beham, A., Kommenda, M., Scheibenpflug, A., Pitzer, E., Vonolfen, S., Kofler, M., Winkler, S.M., Dorfer, V., Affenzeller, M.: Architecture and design of the heuristiclab optimization environment. Advanced Methods and Applications in Computational Intelligence, Topics in Intelligent Engineering and Informatics **6**, 197–261 (2013)
24. White, D.: An overview of schema theory. Computing Research Repository CoRR **abs/1401.2651** (2014). http://arxiv.org/abs/1401.2651
25. Woodward, J.R.: Modularity in Genetic Programming. Proc. of Genetic Programming: 6th European Conference, EuroGP 2003 Essex, pp. 254–263. Springer (2003). http://dx.doi.org/10.1007/3-540-36599-0_23

Chapter 3
Genetic Programming Symbolic Classification: A Study

Michael F. Korns

Abstract While Symbolic Regression (SR) is a well-known offshoot of Genetic Programming, Symbolic Classification (SC), by comparison, has received only meager attention. Clearly, regression is only half of the solution. Classification also plays an important role in any well rounded predictive analysis tool kit. In several recent papers, SR algorithms are developed which move SR into the ranks of extreme accuracy. In an additional set of papers algorithms are developed designed to push SC to the level of basic classification accuracy competitive with existing commercially available classification tools. This paper is a simple study of four proposed SC algorithms and five well-known commercially available classification algorithms to determine just where SC now ranks in competitive comparison. The four SC algorithms are: simple genetic programming using argmax referred to herein as (AMAXSC); the M_2GP algorithm; the MDC algorithm, and Linear Discriminant Analysis (LDA). The five commercially available classification algorithms are available in the KNIME system, and are as follows: Decision Tree Learner (DTL); Gradient Boosted Trees Learner (GBTL); Multiple Layer Perceptron Learner (MLP); Random Forest Learner (RFL); and Tree Ensemble Learner (TEL). A set of ten artificial classification problems are constructed with no noise. The simple formulas for these ten artificial problems are listed herein. The problems vary from linear to nonlinear multimodal and from 25 to 1000 columns. All problems have 5000 training points and a separate 5000 testing points. The scores, on the out of sample testing data, for each of the nine classification algorithms are published herein.

M. F. Korns (✉)
Lantern Credit LLC, Henderson, NV, USA

© Springer International Publishing AG, part of Springer Nature 2018
W. Banzhaf et al. (eds.), *Genetic Programming Theory and Practice XV*,
Genetic and Evolutionary Computation, https://doi.org/10.1007/978-3-319-90512-9_3

3.1   Introduction

Symbolic Regression (SR) is a well-known offshoot of Genetic Programming; however, Symbolic Classification (SC) by comparison, has received relatively little attention. While regression is important, it is only half of the solution. Classification also plays an important role in any well rounded predictive analysis tool kit. Several recent papers develop algorithms which move SR into the ranks of extreme accuracy [2, 3, 5–8]. Additionally several papers develop algorithms designed to raise SC accuracy to the level of basically competitive with existing commercially available classification tools [1, 9, 10, 14].

This paper is a simple study of four proposed SC algorithms and five well-known commercially available classification algorithms to determine just where SC now ranks in competitive comparison. The four SC algorithms are: simple genetic programming using argmax referred to herein as (AMAXSC); the M_2GP algorithm [1]; the MDC algorithm [9], and Linear Discriminant Analysis (LDA) [14]. The five commercially available classification algorithms are available in the KNIME system [15], and are as follows: Decision Tree Learner (DTL); Gradient Boosted Trees Learner (GBTL); Multiple Layer Perceptron Learner (MLP); Random Forest Learner (RFL); and Tree Ensemble Learner (TEL).

A set of ten artificial classification problems are constructed with no noise such that absolutely accurate classifications are theoretically possible. The discriminant formulas for these ten artificial problems are listed herein. The problems vary from linear to nonlinear multimodal and from 25 to 1000 columns such that each classification algorithm will be stressed on well understood problems from the simple to the very difficult. All problems have 5000 training points and a separate 5000 testing points. The scores on the out of sample testing data, for each of the nine classification algorithms are published herein.

No assertion is made that these four genetic programming SC algorithms are the best in the literature. In fact we know of an additional enhanced algorithm, *which we have not had time to implement for this study*, M_3GP [10]. No assertion is made that the five KNIME classification algorithms are the best commercially available, only that KNIME is a trusted component of Lantern Credit predictive analytics. This study is simply meant to provide one reference point for how far genetic programming symbolic classification has improved relative to a set of reasonable commercially available classification algorithms.

Each of the four genetic programming SC algorithms is briefly explained further in this paper as follows.

3.2   AMAXSC in Brief

The simplest naive genetic programming approach to multiclass classification is arguably a standard genetic programming approach, such as a modification of the baseline algorithm [4], using the **argmax** function to classify as follows,

$$y = argmax(gp_1, gp_2, \ldots, gp_K) \tag{3.1}$$

where K is the number of classes

Each gp_k represents a separate discriminant function evolved via standard genetic programming. The argmax() function chooses the class (1 to K) which has the highest value, and is strongly related to the Bayesian probability that the training point belongs to the k-th class. No other enhancements are needed other than the standard argmax() function and a slightly modified genetic programming system—modified to evolve one formula for each class instead of the usual single formula.

3.3   MDC in Brief

The Multilayer Discriminant Classification (MDC) algorithm is an evolutionary approach to enhancing the simple AMAXSC algorithm.

$$y = argmax(w_{10} + (w_{11} * gp_1), w_{20} + (w_{21} * gp_2), \ldots, w_{C0} + (w_{C1} * gp_K)) \tag{3.2}$$

where K is the number of classes, the gp_k are GP evolved formulas, and the w_{ij} are real weight coefficients (there are $2K$ weights).

Each gp_k represents a separate discriminant function evolved via standard genetic programming. The argmax() function chooses the class (1 to C) which has the highest value, and is strongly related to the Bayesian probability that the training point belongs to the k-th class. Given a set of GP evolved discriminant formulas $\{gp_1, gp_2, \ldots, gp_K\}$, the objective of the MDC algorithm is to optimize the choice of coefficient weights $\{w_{10}, w_{11}, w_{20}, w_{21}, \ldots, w_{K0}, w_{K1}\}$ so that Eq. (3.2) is optimized for all X and Y.

The first step in the MDC algorithm is to perform a Partial Bipolar Regression on each discriminant entry i.e. $w_{k0} + (w_{k1} \times gp_k) = Y_k + e$. This produces starting weights for w_{k0} and w_{k1} which are not very good but are much better than random. The second step in the MDC algorithm is to run a Modified Sequential Minimization on selected discriminant entries. This produces much better weight candidates for all discriminant functions, but still not perfect. Finally, the MDC algorithm employs the Bees Algorithm to fully optimize the coefficient weights.

The MDC algorithm is discussed in much greater detail in [9].

3.4   M₂GP in Brief

The M_2GP algorithm is described in detail in [1]. Briefly the M_2GP algorithm generates a D-dimensional GP tree instead of a 1-dimensional GP tree. Assuming that there are K classes, the algorithm attempts to minimize the Mahalanobis

distance between the n-th training point and the centroid of the k-th class. The basic
training algorithm is as follows.

Algorithm 3.1 (M$_2$GP training)

---

1. Input: X, Y, D - where X is an $M \times N$ real matrix, Y is an N vector, D is a
   scalar
2. For g from 1 to G do
3. Generate: $F = \{f_1, f_2, \ldots, f_D\}$ set of D solutions
4. Evaluate: $Z_s = \text{Eval}(f_s(X))$ for s from 1 to D – a D-dimensional point
5. Cluster: Z^k in Z for all k from 1 to K – group all the Z which belong to each
   class
6. For k from 1 to C do
7. $C^k = \text{covar}(Z^k)$ – a D by D covariance matrix for each class
8. $W^k = \text{centroid}(Z^k)$ – a 1 by D centroid vector
9. $D_k(X_n) = \text{sqrt}((Z_n - W^k) \times (C^k)^{-1} \times (Z_n - W^k)^T)$ - for n from 1 to N (the
   number of training points)
10. For n from 1 to N do $EY_n = \text{argmin}(D_1(X_n), D_2(X_n), \ldots, D_C(X_n))$
11. For n from 1 to N do $E_n = 1$ IFF $EY_n \neq Y_n$, 0 otherwise
12. Minimize average(EY)
13. Return F, C, M

The M$_2$GP algorithm is discussed in much greater detail in [1].

3.5 LDA Background

Linear Discriminant Analysis (LDA) is a generalization of Fischer's linear discriminant, which is a method to find a linear combination of features which best separates
K classes of training points [11–13]. LDA is used extensively in Statistics, Machine
Learning, and Pattern Recognition.

Similar to the arguments leading up to the M$_2$GP algorithm [1], we argue that any
symbolic regression system can be converted into a symbolic classification system.
In this paper we start with the baseline algorithm published in [4]. Our baseline SR
system inputs an N by M matrix of independent training points, X, and an N vector
of dependent values, Y. The SR system outputs a predictor function, $F(X) \sim Y$
where F is the best least squares estimator for Y which the SR system could find
in the allotted training time. The format of F is important, and consists of one or
more basis functions Bf_b with regression constants c_b. There are always B basis
functions and $B + 1$ coefficients. The following is the format of F.

$$y = c_0 + c_1 * Bf_1 + c_2 * Bf_2 + \cdots + c_B * Bf_B \tag{3.3}$$

There are from 1 to B basis functions with 2 to $B + 1$ real number coefficients. Each basis function is an algebraic combination of operators on the M features of X, such that $Bf_b(X)$ is a real number. The following is a typical example of an SR produced predictor, $F(X)$.

$$y = 2.3 + .9 * cos(x_3) + 7.1 * x_6 + 5.34 * (x_4/tan(x_8)) \tag{3.4}$$

The coefficients c_0 to c_B play an important role in minimizing the least squares error fit of F with Y. The coefficients can be evolved incrementally, but most industrial strength SR systems identify the optimal coefficients via an assisted fitness training technique. In the baseline SR algorithm this assisted fitness training technique is simple linear regression ($B = 1$) or multiple linear regression ($B > 1$). In symbolic classification problems the N by M matrix of independent training points, X, is unchanged. However, The N vector of dependent values contains only categorical unordered values between 1 and K. Furthermore the least squares error fitness measure (LSE) is replaced with classification error percent (CEP) fitness. Therefore we cannot use regression for assisted fitness training in our new SC system. Instead, we can use LDA as an assisted fitness training technique in our new SC system.

Our new SC system now outputs not one predictor function, but instead outputs K predictor functions (one for each class). These functions are called discriminants, $D_k(X) \sim Y_k$, and there is one discriminant function for each class. The format of the SC's discriminant function output is always as follows.

$$y = argmax(D_1, D_2, \ldots, D_K) \tag{3.5}$$

The argmax function returns the class index for the largest valued discriminant function. For instance if $D_i = max(D_1, D_2, \ldots, D_K)$, then $i = argmax(D_1, D_2, \ldots, D_K)$.

A central aspect of LDA is that each discriminant function is a linear variation of every other discriminant function and reminiscent of the multiple basis function estimators output by the SR system. For instance if the GP symbolic classification system produces a candidate with B basis functions, then each discriminant function has the following format.

$$D_0 = c_{00} + c_{01} \times Bf_1 + c_{02} \times Bf_2 + \ldots + c_{0B} \times Bf_B$$
$$D_1 = c_{10} + c_{11} \times Bf_1 + c_{12} \times Bf_2 + \ldots + c_{1B} \times Bf_B \tag{3.6}$$
$$D_2 = c_{20} + c_{21} \times Bf_1 + c_{22} \times Bf_2 + \ldots + c_{2B} \times Bf_B$$

The $K \times (B + 1)$ coefficients are selected so that the i-th discriminant function has the highest value when the $y = i$ (*i.e. the class is i*). The technique for selecting these optimized coefficients c_{00} to c_{KB} is called linear discriminant analysis and in the following section we will present the Bayesian formulas for these discriminant functions.

3.6 LDA Matrix Math

We use Bayes rule to minimize the *classification error percent* (CEP) by assigning a training point $X_{[n]}$ to the class k if the probability of $X_{[n]}$ belonging to class k, $P(k|X_{[n]})$, is higher than the probability for all other classes as follows.

$$EY_{[n]} = k, \text{ iff } P(k|X_{[n]}) \geq P(j|X_{[n]}) \text{ for all } 1 \leq j \geq K \tag{3.7}$$

The CEP is computed as follows.

$$CEP = \sum (EY_{[n]} \neq Y_{[n]}| \text{ for all } n)/N \tag{3.8}$$

Therefore, each discriminant function D_k acts a Bayesian estimated percent probability of class membership in the formula.

$$y = argmax(D_1, D_2, \ldots, D_K) \tag{3.9}$$

The technique of LDA makes three assumptions, (a) that each class has multivariate Normal distribution, (b) that each class covariance is equal, and (c) that the class covariance matrix is nonsingular. Once these assumptions are made, the mathematical formula for the optimal Bayesian discriminant function is as follows.

$$D_k(X_n) = \mu_k(C_k)^{-1}(X_n)^T - 0.5\mu_k(C_k)^{-1}(\mu_k)^T + \ln P_k \tag{3.10}$$

where X_n is the n-th training point, μ_k is the mean vector for the k-th class, $(C_k)^{-1}$ is inverse of the covariance matrix for the k-th class, $(X_n)^T$ is the transpose of the n-th training point, $(\mu_k)^T$ is the transpose of the mean vector for k-th class, and $\ln P_k$ is the natural logarithm of the naive probability that any training point will belong to the k-th class.

In the following section we will present step by step implementation guidelines for LDA assisted fitness training in our new extended baseline SC system, as indicated by the above Bayesian formula for $D_k(X_n)$.

3.7 LDA Assisted Fitness Implementation

The baseline SR system [4] attempts to score thousands to millions of regression candidates in a run. These are presented for scoring via the fitness function which returns the least squares error (LSE) fitness measure.

$$LSE = fitness(X, Y, Bf_1, \ldots, Bf_B, c_0, \ldots, c_B) \tag{3.11}$$

The coefficients $c_0, \ldots, c_B$ can be taken as is, and the simple LSE returned. However, most industrial strength SR systems use regression as an assisted fitness technique to supply optimal values for the coefficients before returning the LSE fitness measure. This greatly speeds up accuracy and allows the SR to concentrate all of its algorithmic resources toward the evolution of an optimal set of basis functions $Bf_1, \ldots, Bf_B$.

Converting to a baseline symbolic classification system will require returning the classification error percent (CEP) fitness measure, *which is defined as the count of erroneous classifications divided by the size of Y*, and extending the coefficients to allow for linear discriminant analysis as follows.

$$CEP = fitness(X, Y, Bf_1, \ldots, Bf_B, c_{00}, \ldots, c_{KB}) \qquad (3.12)$$

Of course the coefficients $c_{00}, \ldots, c_{KB}$ can be taken as is, and the simple CEP returned. However, our new baseline SC system will use LDA as an assisted fitness technique to supply optimal values for the coefficients before returning the CEP fitness measure. This greatly speeds up accuracy and allows the SR to concentrate all of its algorithmic resources toward the evolution of an optimal set of basis functions $Bf_1, \ldots, Bf_B$.

3.7.1 Converting to Basis Space

The first task of our new SC fitness function must be to convert from N by M feature space, X, into N by B basis space XB. Basis space is the training matrix created by assigning basis function conversions to each of the B points in XB as follows.

$$XB_{[n][b]} = Bf_b(X_{[n]}) \qquad (3.13)$$

So for each row n of our N by M input feature space training matrix, $(X_{[n]})$, we apply all B basis functions, yielding the B points of our basis space training matrix for row n, $(XB_{[n][b]})$. Our new training matrix is the N by B basis space matrix, XB.

3.7.2 Class Clusters and Centroids

Next we must compute the K class cluster matrices for each of the K classes as follows: **Define**

$$CX_{[k]} \text{ where } XB_{[n]} \in CX_{[k]} \text{ iff } Y_{[n]} = k \qquad (3.14)$$

Each matrix $CX_{[k]}$ contains only those training rows of XB which belong to the class k. We also compute the simple Bayesian probability of membership in each class cluster matrix as follows.

$$P_{[K]} = length(CX_{[k]})/N \qquad (3.15)$$

Next we compute the K cluster mean vectors, each of which is a vector of length B containing the average value in each of the B columns of each of the K class cluster matrices, CX, as follows.

$$\mu_{[k][b]} = \text{column mean of the b-th column of } CX_{[K]} \qquad (3.16)$$

We next compute the class centroid matrices for each of the K classes, which are simply the mean adjusted class clusters as follows

$$CXU_{[k][m][b]} = (CX_{[k][m][b]} - \mu_{[k][b]}) \text{ for all } k, m, \text{ and } b \qquad (3.17)$$

Finally we must compute the B by B class covariance matrices, which are the class centroid covariance matrices for each class as follows.

$$COV_{[k]} = covarianceMatrix(transpose(CXU_{[k]})) \qquad (3.18)$$

Each of the K class covariance matrices is a B by B covariance matrix for that specified class.

In order to support the core LDA assumption that the class covariance matrices are all equal, we compute the final covariance matrix by combining each class covariance matrix according to their naive Bayesian probabilities as follows. class as follows.

$$C_{[m][n]} = \sum(COV_{[k][m][n]} \times P_{[k]}) \text{ for all } 1 \leqslant k \leqslant K \qquad (3.19)$$

The final treatment of the covariance matrix allows the LDA optimal coefficients to be computed as shown in the following section.

3.7.3 LDA Coefficients

Now we can easily compute the single axis coefficient for each class as follows.

$$c_{[k][0]} = -0.5\mu_k(C_k)^{-1}(\mu_k)^T + \ln P_k \qquad (3.20)$$

The B basis function coefficients for each class are computed as follows.

$$c_{[k][1,...B]} = \mu_k (C_k)^{-1} \tag{3.21}$$

All together these coefficients form the discriminants for each class as follows.

$$y = c_{k0} + c_{k1} * Bf_1(X_n) + c_{k2} * Bf_2(X_n) + \cdots + c_{kB} * Bf_B(X_n) \tag{3.22}$$

And the estimated value for Y is defined as follows.

$$y = argmax(D_1(X_n), D_2(X_n), \ldots, D_K(X_n)) \tag{3.23}$$

3.8 Artificial Test Problems

A set of ten artificial classification problems are constructed, with no noise, to compare the four proposed SC algorithms and five well-known commercially available classification algorithms to determine just where SC now ranks in competitive comparison. The four SC algorithms are: simple genetic programming using argmax referred to herein as (AMAXSC); the M_2GP algorithm [1]; the MDC algorithm [9], and Linear Discriminant Analysis (LDA) [14]. The five commercially available classification algorithms are available in the KNIME system [15], and are as follows: Decision Tree Learner (DTL); Gradient Boosted Trees Learner (GBTL); Multiple Layer Perceptron Learner (MLP); Random Forest Learner (RFL); and Tree Ensemble Learner (TEL).

Each of the artificial test problems is created around an X training matrix filled with random real numbers in the range $[-10.0, +10.0]$. The number of rows and columns in each test problem varies from 5000×25 to 5000×1000 depending upon the difficulty of the problem. The number of classes varies from $Y = 1, 2$ to $Y = 1, 2, 3, 4, 5$ depending upon the difficulty of the problem. The test problems are designed to vary from extremely easy to very difficult. The first test problem is linearly separable with two classes on 25 columns. The tenth test problem is nonlinear multimodal with five classes on 1000 columns.

Standard statistical best practices out of sample testing are employed. First training matric X is filled with random real numbers in the range $[-10.0, +10.0]$, and the Y class values are computed from the argmax functions specified below. A champion is trained on the training data. Next a testing matric X is filled with random real numbers in the range $[-10.0, +10.0]$, and the Y class values are computed from the argmax functions specified below. The previously trained champion is run on the testing data and scored against the Y values. Only the out of sample testing scores are shown in the results in Table 3.1.

The argmax functions used to create each of the ten artificial test problems are as follows.

Artificial Test Problems

- T_1: $y = argmax(D_1, D_2)$ where $Y = 1, 2$, X is 5000×25, and each D_i is as follows

$$\begin{cases} D_1 = sum((1.57 * x_0), (-39.34 * x_1), (2.13 * x_2), (46.59 * x_3), (11.54 * x_4)) \\ D_2 = sum((-1.57 * x_0), (39.34 * x_1), (-2.13 * x_2), (-46.59 * x_3), (-11.54 * x_4)) \end{cases}$$

- T_2: $y = argmax(D_1, D_2)$ where $Y = 1, 2$, X is 5000×100, and each D_i is as follows

$$\begin{cases} D_1 = sum((5.16 * x_0), (-19.83 * x_1), (19.83 * x_2), (29.31 * x_3), (5.29 * x_4)) \\ D_2 = sum((-5.16 * x_0), (19.83 * x_1), (-0.93 * x_2), (-29.31 * x_3), (5.29 * x_4)) \end{cases}$$

- T_3: $y = argmax(D_1, D_2)$ where $Y = 1, 2$, X is 5000×1000, and each D_i is as follows

$$\begin{cases} D_1 = sum((-34.16 * x_0), (2.19 * x_1), (-12.73 * x_2), (5.62 * x_3), (-16.36 * x_4)) \\ D_2 = sum((34.16 * x_0), (-2.19 * x_1), (12.73 * x_2), (-5.62 * x_3), (16.36 * x_4)) \end{cases}$$

- T_4: $y = argmax(D_1, D_2, D_3)$ where $Y = 1, 2, 3$, X is 5000×25, and each D_i is as follows

$$\begin{cases} D_1 = sum((1.57 * cos(x_0)), (-39.34 * square(x_{10})), (2.13 * (x_2/x_3)), \\ \quad (46.59 * cube(x_{13})), (-11.54 * log(x_4))) \\ D_2 = sum((-0.56 * cos(x_0)), (9.34 * square(x_{10})), (5.28 * (x_2/x_3)), \\ \quad (-6.10 * cube(x_{13})), (1.48 * log(x_4))) \\ D_3 = sum((1.37 * cos(x_0)), (3.62 * square(x_{10})), (4.04 * (x_2/x_3)), \\ \quad (1.95 * cube(x_{13})), (9.54 * log(x_4))) \end{cases}$$

- T_5: $y = argmax(D_1, D_2, D_3)$ where $Y = 1, 2, 3$, X is 5000×100, and each D_i is as follows

$$\begin{cases} D_1 = sum((1.57 * sin(x_0)), (-39.34 * square(x_{10})), (2.13 * (x_2/x_3)), \\ \quad (46.59 * cube(x_{13})), (-11.54 * log(x_4))) \\ D_2 = sum((-0.56 * sin(x_0)), (9.34 * square(x_{10})), (5.28 * (x_2/x_3)), \\ \quad (-6.10 * cube(x_{13})), (1.48 * log(x_4))) \\ D_3 = sum((1.37 * sin(x_0)), (3.62 * square(x_{10})), (4.04 * (x_2/x_3)), \\ \quad (1.95 * cube(x_{13})), (9.54 * log(x_4))) \end{cases}$$

- T_6: $y = argmax(D_1, D_2, D_3)$ where $Y = 1, 2, 3$, X is 5000×1000, and each D_i is as follows

$$
\begin{cases}
D_1 = sum((1.57 * tanh(x_0)), (-39.34 * square(x_{10})), (2.13 * (x_2/x_3)), \\
(46.59 * cube(x_{13})), (-11.54 * log(x_4))) \\
D_2 = sum((-0.56 * tanh(x_0)), (9.34 * square(x_{10})), (5.28 * (x_2/x_3)), \\
(-6.10 * cube(x_{13})), (1.48 * log(x_4))) \\
D_3 = sum((1.37 * tanh(x_0)), (3.62 * square(x_{10})), (4.04 * (x_2/x_3)), \\
(1.95 * cube(x_{13})), (9.54 * log(x_4)))
\end{cases}
$$

- T_7: $y = argmax(D_1, D_2, D_3, D_4, D_5)$ where $Y = 1, 2, 3, 4, 5$, X is 5000×25, and each D_i is as follows

$$
\begin{cases}
D_1 = sum((1.57 * cos(x_0/x_{21})), (9.34 * ((square(x_{10})/x_{14}) * x_6)), \\
(2.13 * ((x_2/x_3) * log(x_8))), (46.59 * (cube(x_{13}) * (x_9/x_2))), \\
(-11.54 * log(x_4 * x_{10} * x_{15}))) \\
D_2 = sum((-1.56 * cos(x_0/x_{21})), (7.34 * ((square(x_{10})/x_{14}) * x_6)), \\
(5.28 * ((x_2/x_3) * log(x_8))), (-6.10 * (cube(x_{13}) * (x_9/x_2))), \\
(1.48 * log(x_4 * x_{10} * x_{15}))) \\
D_3 = sum((2.31 * cos(x_0/x_{21})), (12.34 * ((square(x_{10})/x_{14}) * x_6)), \\
(-1.28 * ((x_2/x_3) * log(x_8))), (0.21 * (cube(x_{13}) * (x_9/x_2))), \\
(2.61 * log(x_4 * x_{10} * x_{15}))) \\
D_4 = sum((-0.56 * cos(x_0/x_{21})), (8.34 * ((square(x_{10})/x_{14}) * x_6)), \\
(16.71 * ((x_2/x_3) * log(x_8))), (-2.93 * (cube(x_{13}) * (x_9/x_2))), \\
(5.228 * log(x_4 * x_{10} * x_{15}))) \\
D_5 = sum((1.07 * cos(x_0/x_{21})), (-1.62 * ((square(x_{10})/x_{14}) * x_6)), \\
(-0.04 * ((x_2/x_3) * log(x_8))), (-0.95 * (cube(x_{13}) * (x_9/x_2))), \\
(0.54 * log(x_4 * x_{10} * x_{15})))
\end{cases}
$$

- T_8: $y = argmax(D_1, D_2, D_3, D_4, D_5)$ where $Y = 1, 2, 3, 4, 5$, X is 5000×100, and each D_i is as follows

$$\begin{cases} D_1 = sum((1.57 * sin(x_0/x_{11})), (9.34 * ((square(x_{12})/x_4) * x_{46})), \\ (2.13 * ((x_{21}/x_3) * log(x_{18}))), (46.59 * (cube(x_3) * (x_9/x_2))), \\ (-11.54 * log(x_{14} * x_{10} * x_{15}))) \\ D_2 = sum((-1.56 * sin(x_0/x_{11})), (7.34 * ((square(x_{12})/x_4) * x_{46})), \\ (5.28 * ((x_{21}/x_3) * log(x_{18}))), (-6.10 * (cube(x_3) * (x_9/x_2))), \\ (1.48 * log(x_{14} * x_{10} * x_{15}))) \\ D_3 = sum((2.31 * sin(x_0/x_{11})), (12.34 * ((square(x_{12})/x_4) * x_{46})), \\ (-1.28 * ((x_{21}/x_3) * log(x_{18}))), (0.21 * (cube(x_3) * (x_9/x_2))), \\ (2.61 * log(x_{14} * x_{10} * x_{15}))) \\ D_4 = sum((-0.56 * sin(x_0/x_{11})), (8.34 * ((square(x_{12})/x_4) * x_{46})), \\ (16.71 * ((x_{21}/x_3) * log(x_{18}))), (-2.93 * (cube(x_3) * (x_9/x_2))), \\ (5.228 * log(x_{14} * x_{10} * x_{15}))) \\ D_5 = sum((1.07 * sin(x_0/x_{11})), (-1.62 * ((square(x_{12})/x_4) * x_{46})), \\ (-0.04 * ((x_{21}/x_3) * log(x_{18}))), (-0.95 * (cube(x_3) * (x_9/x_2))), \\ (0.54 * log(x_{14} * x_{10} * x_{15}))) \end{cases}$$

- T_9: $y = argmax(D_1, D_2, D_3, D_4, D_5)$ where $Y = 1, 2, 3, 4, 5$, X is 5000×1000, and each D_i is as follows

$$\begin{cases} D_1 = sum((1.57 * sin(x_{20} * x_{11})), (9.34 * (tanh(x_{12}/x_4) * x_{46})), \\ (2.13 * ((x_{321} - x_3) * tan(x_{18}))), (46.59 * (square(x_3)/(x_{49} * x_{672}))), \\ (-11.54 * log(x_{24} * x_{120} * x_{925}))) \\ D_2 = sum(((-1.56) * sin(x_{20} * x_{11})), (7.34 * (tanh(x_{12}/x_4) * x_{46})), \\ (5.28 * ((x_{321} - x_3) * tan(x_{18}))), ((-6.10) * (square(x_3)/(x_{49} * x_{672}))), \\ (1.48 * log(x_{24} * x_{120} * x_{925}))) \\ D_3 = sum((2.31 * sin(x_{20} * x_{11})), (12.34 * (tanh(x_{12}/x_4) * x_{46})), \\ ((-1.28) * ((x_{321} - x_3) * tan(x_{18}))), (0.21 * (square(x_3)/(x_{49} * x_{672}))), \\ (2.61 * log(x_{24} * x_{120} * x_{925}))) \\ D_4 = sum(((-0.56) * sin(x_{20} * x_{11})), (8.34 * (tanh(x_{12}/x_4) * x_{46})), \\ (16.71 * ((x_{321} - x_3) * tan(x_{18}))), ((-2.93) * (square(x_3)/(x_{49} * x_{672}))), \\ (5.228 * log(x_{24} * x_{120} * x_{925}))) \\ D_5 = sum((1.07 * sin(x_{20} * x_{11})), ((-1.62) * (tanh(x_{12}/x_4) * x_{46})), \\ ((-0.04) * ((x_{321} - x_3) * tan(x_{18}))), ((-0.95) * (square(x_3)/(x_{49} * x_{672}))), \\ (0.54 * log(x_{24} * x_{120} * x_{925}))) \end{cases}$$

- T_{10}: $y = argmax(D_1, D_2, D_3, D_4, D_5)$ where $Y = 1, 2, 3, 4, 5$, X is 5000×1000, and each D_i is as follows

$$
\begin{cases}
D_1 = sum((1.57 * sin(x_{20} * x_{11})), (9.34 * (tanh(x_{12}/x_4) * x_{46})), \\
\quad (2.13 * ((x_{321} - x_3) * tan(x_{18}))), (46.59 * (square(x_3)/(x_{49} * x_{672}))), \\
\quad (-11.54 * log(x_{24} * x_{120} * x_{925}))) \\[4pt]
D_2 = sum(((-1.56) * sin(x_{20} * x_{11})), (7.34 * (tanh(x_{12}/x_4) * x_{46})), \\
\quad (5.28 * ((x_{321} - x_3) * tan(x_{18}))), ((-6.10) * (square(x_3)/(x_{49} * x_{672}))), \\
\quad (1.48 * log(x_{24} * x_{120} * x_{925}))) \\[4pt]
D_3 = sum((2.31 * sin(x_{20} * x_{11})), (12.34 * (tanh(x_{12}/x_4) * x_{46})), \\
\quad ((-1.28) * ((x_{321} - x_3) * tan(x_{18}))), (0.21 * (square(x_3)/(x_{49} * x_{672}))), \\
\quad (2.61 * log(x_{24} * x_{120} * x_{925}))) \\[4pt]
D_4 = sum(((-0.56) * sin(x_{20} * x_{11})), (8.34 * (tanh(x_{12}/x_4) * x_{46})), \\
\quad (16.71 * ((x_{321} - x_3) * tan(x_{18}))), ((-2.93) * (square(x_3)/(x_{49} * x_{672}))), \\
\quad (5.228 * log(x_{24} * x_{120} * x_{925}))) \\[4pt]
D_5 = sum((1.07 * sin(x_{20} * x_{11})), ((-1.62) * (tanh(x_{12}/x_4) * x_{46})), \\
\quad ((-0.04) * ((x_{321} - x_3) * tan(x_{18}))), ((-0.95) * (square(x_3)/(x_{49} * x_{672}))), \\
\quad (0.54 * log(x_{24} * x_{120} * x_{925})))
\end{cases}
$$

The Symbolic Classification system for all four SC algorithms (AMAXSC, M_2GP, MDC, LDA) avail themselves of the following operators:

- Binary Operators: + - / × minimum maximum
- Relational Operators: $<, \leq, =, \neq, \geq, >$
- Logical Operators: lif land lor
- Unary Operators: inv abs sqroot square cube curoot quart quroot exp ln binary sign sig cos sin tan tanh

The unary operators sqroot, curoot, and quroot are square root, cube root, and quart root respectively. The unary operators inv, ln, and sig are the inverse, natural log, and sigmoid functions respectively.

3.9   Performance on Test Problems

Here we compare the out of sample CEP testing scores of the four proposed SC algorithms and five well-known commercially available classification algorithms to determine where SC ranks in competitive comparison. The four SC algorithms are: simple genetic programming using argmax referred to herein as (AMAXSC); the M$_2$GP algorithm; the MDC algorithm, and Linear Discriminant Analysis (LDA). The five commercially available classification algorithms are available in the KNIME system, and are as follows: Decision Tree Learner (DTL); Gradient Boosted Trees Learner (GBTL); Multiple Layer Perceptron Learner (MLP); Random Forest Learner (RFL); and Tree Ensemble Learner (TEL). On a positive note, the three new proposed symbolic classifications algorithms are a great improvement over the vanilla AMAXSC algorithm. On a negative note, none of the three newly proposed SC algorithms are the best performer. The top performer overall by a good margin is the Gradient Boosted Trees Learner (GBTL).

It is interesting to note that all three newly proposed SC algorithms performed better overall than the Multiple Layer Perceptron Learner (MLP). This is significant; as it is the first time we have seen a genetic programming SC algorithm beat one of the top commercially available classification algorithms.

Of the three newly proposed SC algorithms, surprisingly the MDC algorithm was the best overall performer; although all three SC algorithms were grouped very close together in performance.

Clearly progress has been made in the development of commercially competitive SC algorithms. But, a great deal more work has to be done before SC can outperform the Gradient Boosted Trees Learner (GBTL).

Table 3.1 Test problem CEP testing results

Test	AMAXSC	LDA	M$_2$GP	MDC	DTL	GBTL	MLP	RFL	TEL
T_1	0.0808	0.0174	0.0330	0.0330	0.0724	0.0308	0.0072	0.0492	0.0496
T_2	0.1108	0.0234	0.0656	0.0402	0.0740	0.0240	0.0360	0.0664	0.0648
T_3	0.1436	0.0182	0.1010	0.0774	0.0972	0.0332	0.0724	0.1522	0.1526
T_4	0.1954	0.0188	0.0180	0.0162	0.0174	0.0170	0.0472	0.0260	0.0252
T_5	0.1874	0.1026	0.1052	0.1156	0.0858	0.0530	0.3250	0.0920	0.0946
T_6	0.6702	0.5400	0.4604	0.5594	0.5396	0.3198	0.6166	0.6286	0.6284
T_7	0.4466	0.4002	0.4060	0.4104	0.2834	0.2356	0.4598	0.2292	0.2284
T_8	0.8908	0.4176	0.4006	0.4124	0.2956	0.2340	0.4262	0.2250	0.2248
T_9	0.8236	0.7450	0.7686	0.6210	0.6058	0.4286	0.6904	0.4344	0.4334
T_{10}	0.8130	0.7562	0.7330	0.6440	0.5966	0.4286	0.5966	0.4296	0.4352
Avg	0.4362	0.3039	0.3091	0.2930	0.2668	0.1805	0.3277	0.2333	0.2337

3.10 Conclusion

Several papers have proposed GP Symbolic Classification algorithms for multi-class classification problems [1, 9, 10, 14]. Comparing these newly proposed SC algorithms with the performance of five commercially available classifications algorithms shows that progress has been made. The three newly proposed SC algorithms now outperform one of the top commercially available algorithms on a set of artificial test problems of varying degrees of difficulty.

It may be significant that, of the commercially available classifiers, the best overall performers are all tree learners. While the random forest learner (RFL) and the tree ensemble learner (TEL) enjoy enhanced performance over the decision tree learner (DTL), it is the gradient boosted tree learner (GBTL) which clearly enjoys the overall top performer slot on these ten artificial test problems.

It will be interesting to see if gradient boosting can be adapted to the three newly proposed SC algorithms (M_2GP, MDC, and LDA) such that they will also enjoy enhanced performance.

Acknowledgements Our thanks to: Thomas May from Lantern Credit for assisting with the KNIME Learner training/scoring on all ten artificial classification problems.

References

1. Ingalalli, Vijay, Silva, Sara, Castelli, Mauro, Vanneschi, Leonardo 2014. *A Multi-dimensional Genetic Programming Approach for Multi-class Classification Problems*. Euro GP 2014 Springer, pp. 48–60.
2. Korns, Michael F. 2013. *Extreme Accuracy in Symbolic Regression*. Genetic Programming Theory and Practice XI. Springer, New York, NY, pp. 1–30.
3. Koza, John R. 1992. *Genetic Programming: On the Programming of Computers by means of Natural Selection*. The MIT Press. Cambridge, Massachusetts.
4. Korns, Michael F. 2012. *A Baseline Symbolic Regression Algorithm*. Genetic Programming Theory and Practice X. Springer, New York, NY.
5. Keijzer, Maarten. 2003. *Improving Symbolic Regression with Interval Arithmetic and Linear Scaling*. European Conference on Genetic Programming. Springer, Berlin, pp. 275–299.
6. Billard, Billard., Diday, Edwin. 2003. *Symbolic Regression Analysis*. Springer. New York, NY.
7. Korns, Michael F. 2015. *Extremely Accurate Symbolic Regression for Large Feature Problems*. Genetic Programming Theory and Practice XII. Springer, New York, NY, pp. 109–131.
8. Korns, Michael F. 2016. *Highly Accurate Symbolic Regression for Noisy Training Data*. Genetic Programming Theory and Practice XIII. Springer, New York, NY, pp. 91–115.
9. Korns, Michael F. 2018. *An Evolutionary Algorithm for Big Data Multi-class Classification Problems*. In William Tozier and Brian W. Goldman and Bill Worzel and Rick Riolo *editors*, Genetic Programming Theory and Practice XIV, Ann Arbor, USA, 2016. www.cs.bham.ac.uk/~wbl/biblio/gp-html/MichaelKorns.html.
10. Munoz, Louis, Silva, Sara, M. Castelli, Trujillo 2014. *M_3GP Multiclass Classification with GP*. Proceedings Euro GP 2015 Springer, pp. 78–91.
11. Fisher, R. A. 1936. *The Use of Multiple Measurements in Taxonomic Problems*. Annals of Eugenics 7 (2) 179–188.

12. Friedman, J. H. 1989. *Regularized Discriminant Analysis*. Journal of American Statistical Association 84 (405) 165–175.
13. McLachan, Geoffrey, J. 2004. *Discriminant Analysis and Statistical Pattern Recognition*. Wiley. New York, NY.
14. Korns, Michael F., 2017. *Evolutionary Linear Discriminant Analysis for Multiclass Classification Problems*. GECCO Conference Proceedings '17, July 15–19, Berlin, Germany. ACM Press, New York (2017), pp. 233–234.
15. Michael R. Berthold, Nicolas Cebron, Fabian Dill, Thomas R. Gabriel, Tobias Kötter, Thorsten Meinl, Peter Ohl, Christoph Sieb, Kilian Thiel, and Bernd Wiswedel, 2007. *KNIME: The Konstanz Information Miner*. ACM SIGKDD Explorations Newsletter. ACM Press, New York (2009), pp. 26–31.

Chapter 4
Problem Driven Machine Learning by Co-evolving Genetic Programming Trees and Rules in a Learning Classifier System

Ryan J. Urbanowicz, Ben Yang, and Jason H. Moore

Abstract A persistent challenge in data mining involves matching an applicable as well as effective machine learner to a target problem. One approach to facilitate this process is to develop algorithms that avoid modeling assumptions and seek to adapt to the problem at hand. Learning classifier systems (LCSs) have proven themselves to be a flexible, interpretable, and powerful approach to classification problems. They are particularly advantageous with respect to multivariate, complex, or heterogeneous patterns of association. While LCSs have been successfully adapted to handle continuous-valued endpoint (i.e. regression) problems, there are still some key performance deficits with respect to model prediction accuracy and simplicity when compared to other machine learners. In the present study we propose a strategy towards improving LCS performance on supervised learning continuous-valued endpoint problems. Specifically, we hypothesize that if an LCS population includes and co-evolves two disparate representations (i.e. LCS rules, and genetic programming trees) than the system can adapt the appropriate representation to best capture meaningful patterns of association, regardless of the complexity of that association, or the nature of the endpoint (i.e. discrete vs. continuous). To successfully integrate these modeling representations, we rely on multi-objective fitness (i.e. accuracy, and instance coverage) and an information exchange mechanism between the two representation 'species'. This paper lays out the reasoning for this approach, introduces the proposed methodology, and presents basic preliminary results supporting the potential of this approach as an area for further evaluation and development.

R. J. Urbanowicz (✉) · J. H. Moore
Institute for Biomedical Informatics, Perelman School of Medicine, University of Pennsylvania, Philadelphia, PA, USA
e-mail: ryanurb@upenn.edu; jhmoore@upenn.edu

B. Yang
Institute for Biomedical Informatics, University of Pennsylvania, Philadelphia, PA, USA
e-mail: yangben@sas.upenn.edu

© Springer International Publishing AG, part of Springer Nature 2018
W. Banzhaf et al. (eds.), *Genetic Programming Theory and Practice XV*,
Genetic and Evolutionary Computation, https://doi.org/10.1007/978-3-319-90512-9_4

4.1   Introduction

Learning classifier systems (LCSs) challenge the typical paradigm of modeling
in classification and regression [22] by evolving a piecewise, distributed model
comprised of rules. These rule-based machine learning techniques are flexible and
powerful largely because they avoid making assumptions about the underlying pat-
tern(s) of association in a given dataset. While LCSs have repeatedly demonstrated
their efficacy on complex problems, e.g. detecting epistatic interactions between
feature combinations [2, 15, 23], one of the more unique abilities of rule-based
machine learning is their ability to model and characterize heterogeneous patterns of
association where different features or subsets of features are predictive in different
respective subsets of data instances [14, 21, 24]. Solving these types of problems
effectively with traditional machine learning approaches would require successful
dataset stratification prior to modeling. Unfortunately, proper stratification is often
not possible without reliable prior knowledge of the underlying heterogeneity.

To date, the vast majority of LCS research has focused on reinforcement learning
problems and the development of LCS strategies designed to address these types
of problems [22]. An emerging focus over the last decade, seeks to adapt and
optimize these rule-based systems to the particulars of supervised learning, where
the endpoint label is always provided during training [2, 3, 11, 19]. We use the
term 'endpoint' to generically refer to the dependent variable, also referred to
as the 'class' or 'action'. Previously, we developed, an LCS called the Extended
Supervised Tracking and Classifying System (ExSTraCS), specifically designed to
facilitate the detection and characterization of complex, epistatic, and heterogeneous
patterns of association in bioinformatics supervised learning problems [19, 24].
Bioinformatics problems, particularly those derived from the field of genomics, are
characteristically noisy and often involve a large number of potentially predictive
features (discrete or continuous-valued). The ExSTraCS algorithm and it's algorithmic predecessors have been successful in solving complex classification problems,
including simulated genomic data embedded with concurrent patters of epistasis and
genetic heterogeneity [15, 19], real-world genetic data with similar patterns [18],
and most recently, the first and still only algorithm to report directly solving the
135-bit multiplexer problem [24]. The multiplexer family of problems, including
the lower-order 6-bit and 11-bit variations, represent a traditional and scalable set
of benchmarks in the evolutionary algorithm community [20]. Solving the 135-bit
multiplexer problem requires identifying 128 distinct (i.e. heterogeneous) subgroups
of features that are each epistatically associated with endpoint within the subset of
data instances they are relevant to.

LCSs have been advantageous for solving extremely complex classification
problems, however, when dealing with simple classification problems or regression
problems (i.e. continuous endpoints or quantitative traits) there can be distinct
disadvantages in adopting the distributed modeling paradigm of rule-based machine
learning. Specifically, (1) rule-based systems can require a large number of rules
to cover the problem space of simpler non-heterogeneous problems, where a classic

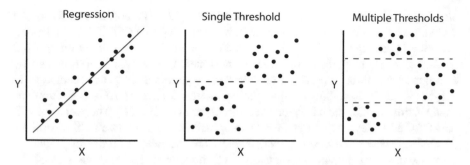

Fig. 4.1 Illustration of continuous-valued endpoint scenarios

model such as a genetic programming tree, might be able to cover that same problem just as accurately with a single tree. Additionally, (2) rule-based systems that learn continuous endpoint intervals as actions, such as ExSTraCS [17], struggle to make precise endpoint predictions in regression problems. However, notably this interval approach is advantages when it comes to other continuous-valued endpoint problem scenarios.

Figure 4.1 identifies three example scenarios of continuous endpoint problems where the independent variable 'X' has a meaningful predictive relationship with the dependent variable 'Y'. First there is a traditional regression scenario where we use the value of X to try and predict a particular value of Y while minimizing error. This assumes that there is an accurate, mapping from independent variable(s) to a precise continuous-valued endpoint. The next two scenarios illustrate what we are calling *threshold models* [17], where the continuous variable Y could more effectively be binned into categorical groups for classification rather than regression. In other words there are meaningful underlying discretizations of the dependent variable range. Thus, the precise predicted value of Y is not particularly informative, but rather it is informative that Y is above, below, or between some 'threshold(s)' illustrated by the dotted red lines in Fig. 4.1. The problem here is that in many real world situations it is unlikely that a practitioner would know ahead of time when binning the endpoint is relevant, how many bins to use, and where to place the thresholds. An algorithm like ExSTraCS could resolve these issues automatically. These threshold-based dependent variable problems are worth considering, for example, in the context of biological gene regulation where some minimum quantity of a repressor or inducer element triggers or prevents some secondary event.

Beyond ExSTraCS, previous efforts to adapt LCS algorithms to continuous endpoint problems have focused on traditional regression or function approximation problems (i.e. no threshold models). Further, efforts have focused on systems designed for reinforcement learning (RL), such as those attempting to control movement of robotics components over a continuous range of angles [5], rather than supervised learning (SL).

The first RL-based LCSs to produce real-valued outputs utilized fuzzy logic [4, 26], applied as controllers for continuous output assignment, or to determine the degree of membership among discrete classes [12]. The best known RL-based LCS algorithm to date, named XCS [27], was extended to XCSF [29, 30] adapting its reward prediction mechanism to calculate a prediction rather than rely on fixed scalars in order to perform function approximation. This was later followed by a proposal for three separate approaches for continuous endpoint prediction in RL-based LCSs including a Interpolating Action Listener (IAL), a Continuous Actor-Critic (CAC), and a General Classifier System (GCS) [31]. The proposed IAL and CAC add a layer of computational complexity by running two LCS algorithms in tandem, where one observes and learns from the other in order to produce a continuous action output. Alternatively GCS integrated the action as part of the rule condition, and learned a simple interval predicate capturing a rang of action values. An extension of XCSF, called XCS-FCA, added an action weight vector to compute real-valued actions along with an evolutionary update of these action weights. Increasing the complexity of rule representation, XCSRCFA and XCSCFA expanded upon XCSR [28] and XCS [27] respectively, replacing discrete actions with a code fragment representation [7, 8]. These code fragments are tree-based expressions similar to those found in genetic programming.

While many of these RL-based LCS approaches to continuous-valued endpoint prediction offer useful ideas, they are not directly translatable to supervised learning systems such as ExSTraCS. For example, the SL algorithm, ExSTraCS, has introduced some uniquely effective mechanisms such as 'attribute tracking' [14] that rely on the formation of a correct set, something that does not exist in RL-based LCSs. Additionally, these methods do not address the interpretability of the resulting rule population, which is of significant import in applications such as bioinformatics, or data mining in general. Furthermore, each of these previous approaches has utilized a more sophisticated representation (e.g. like a genetic programming tree) as part of the condition or action of a rule. In other words the learning entity is still a rule, that must have a satisfied condition to be relevant to a given training instance. Previously, a few other attempts have been made to utilize GP-like tree structures into LCS rule conditions [1, 10, 13]. However these works substitute the classic LCS rule condition with a tree, rather than integrating traditional rules with traditional trees as complete model candidates. To the best of our knowledge the combination of rules and GP trees as separate co-evolving modeling entities has yet to be explored.

As mentioned, in previous work we proposed an extension to the supervised learning ExSTraCS LCS algorithm considering all of the continuous endpoint scenarios described above [17]. This extension incorporated endpoint intervals into rule actions (as opposed to a static class value), as well as a novel prediction scheme that converted these intervals into a specific continuous-value prediction. That work revealed that rules with interval actions retained the ability to detect heterogeneous associations in regression problems and displayed the unique ability to successfully and automatically model the threshold-based continuous endpoint problems. However, with regards to traditional regression geared at predicting precise continuous values, it was found that while the predictive features could still be identified, the prediction error of these rule-based models was much larger than would be expected for traditional modelling approaches.

In this paper, we propose an integrated strategy to deal with these problem-specific rule-based machine learning shortcomings. In recognizing that for some problems, a single model may be preferable to a distributed rule-based model, we seek to merge the best of both worlds into a single learning system. In doing so we seek to retain, if not enhance, the functionality and interpretability of the original ExSTraCS system. Therefore we propose to integrate genetic programming (GP) tree modeling as a co-evolving alternative to the 'condition:action' rule representation of an LCS. We consider these two different representations to be separate species evolving within a finite population managed by the LCS. This is because rules and trees have distinctly different architectures unable to be crossed with one another to form offspring in the same way that can be achieved in evolutionary systems with a uniform representation (see Table 4.1). We have focused on integrating GP trees since they have classically been successful in accurately modeling regression problems [9], and are widely recognized as a largely interpretable approach to modeling in contrast with methods such as artificial neural networks or random forests [6]. In addition to the algorithm automatically choosing the most appropriate model representation, we consider that there may be an inherent learning advantage in using duel representations, such that useful building blocks of information (particularly with respect to informative features and feature combinations) may be more easily captured by one representation or the other depending on the problem, which could then be transferred across representations, thereby accelerating learning.

Table 4.1 Comparing LCS rules and GP trees

LCS rule (ExSTraCS-style representation [24])
• Can only cover (i.e. is relevant to) instances that it matches.
• Does not use functions or operators.
• Features (and a specific value or range) are either specified or instead generalized with a 'don't care' (i.e. #).
• Rule population starts off empty and relies on covering for smart initialization.
• Evolutionary operators: mutation and crossover—can specialize or generalize features.

GP tree (traditional representation)
• Covers all instances.
• Relies on functions and operators.
• Features are treated as variables with unspecified values.
• Tree population is randomly initialized.
• Evolutionary operators: mutation and crossover—can manipulate the variables, constants, and operators at nodes or terminals.

Key representation considerations for learning and modeling

Our proposed methodology is unique in its attempt to co-evolve complete models in the form of GP trees, as well as partial models in the form of LCS rules. It is notable that our proposed method is not strictly co-evolution, since we do allow some limited exchange of 'genetic' information between rules and trees, however we believe this exchange is advantageous to the overall learning efficiency. We hypothesize that this methodology will allow the problem to drive the success of either rules or GP trees based on which representation is most effective at capturing the underlying pattern(s) of association. In this preliminary implementation and evaluation, we ask the simple question of whether rules and trees can evolve together in the same population sharing a common multi-objective fitness metric without one entity or the other being eliminated in the modeling process. More importantly does the best suited representation indeed capture the underlying patterns of association.

We believe this proposed approach will be particularly beneficial in problems with unknown patterns of association or to help supervised LCSs function better on regression problems while retaining performance on threshold-style continuous endpoint problems. The rest of this paper will discuss our proposed methodology, and present the preliminary findings associated with this research in progress.

4.2   Methods

In this section we describe (1) ExSTraCS, the LCS algorithm within which we are testing our problem driven co-evolution strategy, (2) our proposed method for integrating GP tree modeling into ExSTraCS, and (3) the datasets and evaluation strategy utilized in preliminary testing.

4.2.1   ExSTraCS

The LCS algorithm utilized in this study is based on the ExSTraCS algorithm detailed in [24]. For a general introduction to LCS algorithms we refer readers to [20, 22]. ExSTraCS is a supervised, Michigan-style learning classifier system descended from the XCS [27] and UCS algorithms [3] that was expanded to better fit the needs of real world problems such as those found in bioinformatics. ExSTraCS focuses on the detection of complex and heterogeneous patters, for both accurate prediction and interpretable knowledge discovery. Specifically the ExSTraCS algorithm combined strategies for improving scalability [24], generating and applying statistically generate expert knowledge for rule initialization and guided evolutionary search [21], and a unique attribute tracking and feedback mechanism for reusing useful building feature combination and explicitly charac-terizing patterns of heterogeneity and candidate heterogeneous subgroups of data instances [14, 18]. Like other Michigan-style LCS algorithms, ExSTraCS evolves a population of individual rules comprised of a 'condition' specifying feature value states (e.g. *IF: featureA = 0 AND featureC = 1*) and an 'action' or 'class' (e.g. *THEN: class = 1*) forming a simple IF:THEN expression that is relevant to the subset of data instances to which the condition applies.

A schematic of the basic ExSTraCS algorithm utilized in this study is given in Fig. 4.2. As a brief summary, ExSTraCS (1) learns incrementally, one training instance at a time from the dataset. (2) This instance is passed to the rule population [P] with a user defined, finite size. In this study, the population can be comprised of either LCS rules, or traditional GP trees collaborating, as well as competing for survival within the restricted population size. (3) Any rule or tree that 'matches', i.e. is relevant to, the current training instance moves on to form a match set [M]. We will refer to rules and trees interchangeably in this study as 'entities'. If we wanted to make predictions on testing data rather than train, at this point any entities in the match set would submit their endpoint prediction to a prediction array, where a voting scheme decides on an overall endpoint prediction based on the collective 'votes' of the matching entities. Going back to a training cycle,

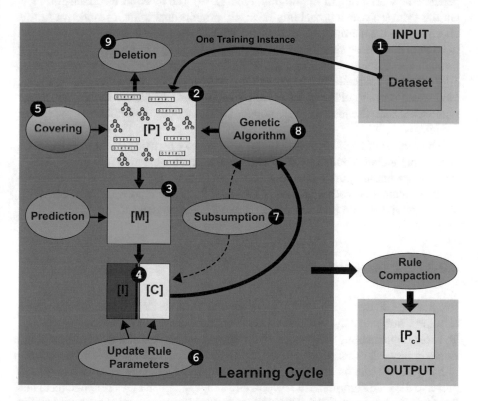

Fig. 4.2 Schematic of the core ExSTraCS algorithm applied in this study. Note that our proposed approach preserves the LCS learning cycle, but integrates GP trees along with rules to comprise the competing and collaborating population of learning entities (i.e. rules and trees)

(4) after the match set is formed, the entities are split into respective 'correct' [C], or 'incorrect' [I] sets based on whether their individual predicted values were correct on the current training instance. (5) At this point if no rules made it to the correct set, the covering mechanism will randomly generate a new rule that both matches and has the correct prediction for the current training instance. Covering is thus responsible for 'intelligently' initializing the rules in the population. (6) Next, the parameters of the entities are updated including their prediction accuracy over all the instances they have matched, and their proportionally related fitness value. Rules that made it to the correct set get an accuracy and fitness boost, while those in the incorrect set receive a loss. (7) Next, the subsumption mechanisms applies an direct generalization pressure to rules in the correct set, where one rule subsumes (i.e. copies itself and replaces another) if it is as accurate and more general (i.e. specifies fewer feature states). Notably in this initial study, subsumption can only operate on rules, not on GP trees. Subsumption is again activated later among parent and offspring rules in the genetic algorithm. (8) The genetic algorithm selects two entities from the correct set proportional to entity fitness. Mutation and crossover operators are applied so that a pair of rules, a pair of trees, or one from each generate new offspring of the original type(s). (9) The deletion mechanism selects entities from [P] with a probability inversely proportional to their respective fitness and removes them from [P] until the population size is less than or equal to the maximum population size. The nine steps of this learning cycle are repeated, cycling through the training dataset for some user defined number of iterations. Lastly, after training, a simple rule compaction mechanism is applied to remove rules (and trees) that are either young (i.e. inexperienced) or have an accuracy below random chance. The resulting rule population constitutes the ExSTraCS prediction model.

The version of ExSTraCS used in this study is an expansion of the one described in [24] that includes both a pareto-front inspired multi-objective rule fitness, rather than the traditional purely accuracy-based rule fitness [25], and that includes the interval-predicate-based expansion of ExSTraCS rules to regression problems [17]. This prototype python implementation is available upon request.

4.2.2 GP Integration

Integrating GP trees into an LCS framework includes a number of challenges resulting from the many differences between these disparate evolutionary systems. To avoid implementing a GP system from the ground up, we utilized the Distributed Evolutionary Algorithms in Python (DEAP) software package for the basic functionality of GP tree initialization, mating, and prediction. Generally, we sought to preserve the LCS algorithm framework, and simply adapt the GP components to this framework wherever needed. Thus, the overall evolutionary cycles are based on the ExSTraCS algorithm rather than a typical GP system. This means that each iteration of the algorithm, the system is exposed to a single training instance, and the system is highly elitist, such that only two new rules or trees are added to the population, with deletion operating to maintain a maximum population size while the rest of the population is preserved.

4.2.2.1   GP Population Initialization

In ExSTraCS and other Michigan-style LCS algorithms, it is typical for the rule population to start out empty, and for the covering operator to activate when needed to initialize rules intelligently that are certain to cover at least the current training instance being learned upon. While this makes sense for rule based systems, GP systems rely on some form of population initialization. Using the GP module in DEAP, trees were allowed to utilize the following operators (addition, subtraction, multiplication, division, negative, negation, less than, greater than, maximum, cosine, and sine), as well as integers ranging from -5 to 5 and all features in the dataset as candidate terminals in the tree. For simplicity in this preliminary study trees were initialized with a depth of two to three. Because DEAP was designed to evolve trees within it's own population structure, we generated DEAP tree populations of size one, and incorporated those into the LCS rule population. In this way, the LCS rule population was initialized with a population half filled with random GP trees, leaving the other half to be filled by rules through covering, and by rule or GP mating.

To give both entities a fair initial balance an equal number of rules were pre-generated with the covering operator before algorithm interactions began. In this study 500 rules and 500 GP trees were generated to initialize the entire entity population.

Notably, LCS covering is typically only activated to generate a new rule when there are currently no rules in the population that both match and make a correct prediction on the current training instance. Since every tree will match every instance, and since it is likely that there will always be at least one tree that makes a correct prediction on the current training instance, we adapted rule covering to activate regardless of the presence of GP trees in the population, but still only when there were no correct or matching rules in the population (as is standard in LCS).

4.2.2.2   GP Parent Selection

The selection of parents for mating by the evolutionary system in ExSTraCS relies on tournament selection as described in [24]. Unlike most co-evolutionary systems, our proposed GP-ExSTraCS algorithm allows for limited mating between GP trees and rules. Therefore parent selection occurs normally within the correct set [C] of the LCS as described in [24]. The correct set is the set of rules (and in this case trees) that both match the current training instance as well as make a correct prediction of endpoint. For discrete endpoints, rules are simply determined to be correct if the action/class component of the rule is equivalent to the class of the instance.

Alternatively, in the case of regression, a rule is included in the correct set if it's interval predicate rage includes the correct endpoint value [17]. For GP trees, the endpoint prediction is not hard coded like the action of a rule, thus the GP is directed to make an endpoint value prediction on the current training instance and this is used to determine if the tree is included in the correct set. This correct set forms the mating pool from which parents can be drawn.

4.2.2.3   GP Mating

If the selected parents are of the same 'species' (i.e. they are both rules or both GP trees) then the discovery of two offspring elements by the genetic algorithm is completed in the standard respective manner. Specifically, rule mating incorporates mutation and crossover as described in [24], and GP tree mating takes place using default DEAP settings and operators.

However if parents of two different species are selected we have introduced here a method that attempts to 'mate' a rule and a tree in the most generic sense. Notably even when a tree and rule are selected as parents mutation operators will still function as they are normally implemented for rules in ExSTraCS and for trees in DEAP. This is not the case for crossover. In particular, since GP tree operators are meaningless in the context of LCS rules, and the specific feature values captured in rules are meaningless to GP trees, the only useful information that can be exchanged is which features are specifically included in either representation. In rules this set of features is captured by the specified attribute list [24], and in trees this is capture by the terminal list (i.e. the set of leaf nodes in the tree), excluding any constants. Next, we identify any features that appear in one representation but not the other. These features are candidates for exchange, each of which has a 50% chance of being exchanged or preserved in the current species. At this there are a few possible scenarios that can occur summarized by Fig. 4.3.

In scenario 1, if there are an equal number of features to be exchanged between rule and tree, we perform an equal swap between species. For example if there is only one feature selected to be exchanged from each species (e.g. feature D in the tree, and feature A in the rule) then feature A will replace feature D in the tree, and feature D (and it's value in the current training instance) will be specified in the rule. Additionally to complete the swap, feature A and it's stored value will be removed from the rule.

In scenario 2, if there is additional feature in the tree selected to be exchanged, then like before that feature will become specified in the rule using the value for the feature in the current training instance. The respective feature terminal is then removed from the tree along with any operator above it in the tree hierarchy.

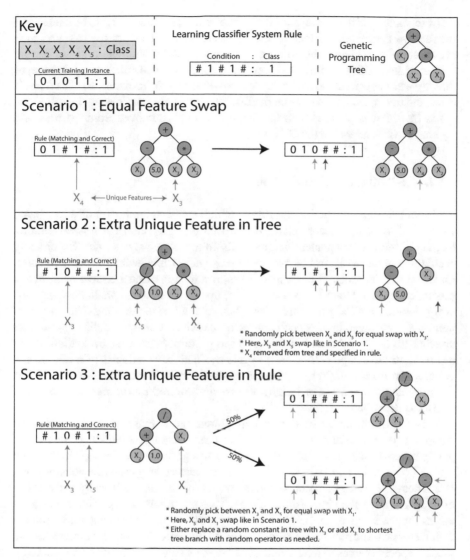

Fig. 4.3 Crossover 'mating' exchange between LCS rules and GP trees. The key illustrates identifies the five 'X' features available in this example, defines a hypothetical current training instance, and offers an example of both and LCS rule and a GP tree. Note that a rule is comprised of two parts, a condition and a predicted class (also referred to as the action). The three scenarios illustrate situations where a rule and tree are chosen as parent entities on the left, and show how information is exchanged to create two new offspring entities (one rule and one tree)

In scenario 3, if there is an additional feature in the rule selected to be exchanged we remove the respective feature and its associated value from the rule and then make one of two choices based on a 50% chance: Add that feature to the tree by replacing an existing terminal that specifies a constant value rather than a feature, or directly add a new terminal to the shortest possible branch (provided that this doesn't cause the tree to exceed maximum depth). If either is not possible we default to the other. If neither is possible then the feature is simply removed from the rule, and nothing more is altered in the GP tree.

4.2.2.4 GP Fitness and Evaluation

Different from traditional GP algorithms which evaluate the fitness of a tree batch wise on all or a significant chunk of the training data, this LCS implementation forces GP trees to be updated incrementally in the same way as rules. Specifically each iteration that a rule or tree matches an instance (which would be every iteration for trees), we determine if the respective representation made a correct or incorrect prediction and keep track of this experience, updating estimated prediction accuracy every iteration until a given rule or tree has seen all of the training data, at which point it is considered to be 'epoch complete' and the accuracy statistic is not longer updated. To allow GP trees in this context to perform batch-wise evaluation would be an unfair advantage with respect to rule evaluation. The fitness of a rule or tree is based on a multi-objective pareto-front-based rule fitness function combining accuracy and the number of instances correctly covered by the rule or tree in the dataset as described in [25].

It is this multi-objective fitness that allows us to reasonably co-evolve and compete rules as partial solutions and trees as candidate full solutions. If fitness was purely based on accuracy, then rules would have a general advantage given that they are not bound to be applicable to all training instances. As a result the accuracy of a rule is better viewed as a 'local' accuracy, where overfit rules will easily reach 100% training accuracy even in noisy problems. This might not be expected of GP trees unless they were given sufficient time and flexibility to evolve deep and complex architectures. If the fitness was purely based on the correct coverage of instances, we would expect GP trees to have a general advantage since they cover all instances by default, thus any reasonably accurate tree may have a much larger correct coverage than the average LCS rule which can only maximally correctly cover the number of instances which it's condition matches. Note that this ExSTraCS multi-objective fitness does not use accuracy and correct coverage directly, but rather surrogate values that only consider accuracy and correct coverage beyond what is expected by random chance. Using these surrogates also helps to account for class imbalance. These surrogate metrics were first introduced to an LCS system in [16] and extended to a pareto-front-based fitness in [25].

4.2.3   Datasets and Evaluation

To initially evaluate our proposed system we selected a handful of simple datasets to explore whether ExSTraCS can adapt itself to apply the most appropriate representation for the underlying problem without prior information. Future work will need to greatly expand the variety of datasets before making any definitive conclusions pertaining to the performance of our proposed GP-LCS co-evolutionary system.

Our initial test datasets focused on classification problems including the 6-bit multiplexer problem described in [24] including six binary attributes and a binary classification. Solving the 6-bit problem requires the identifying four heterogeneous feature combinations each with a 3-way interaction to predict endpoint. Additionally we tested a simple toy dataset with 180 instances and 19 features (with two features additively associated with a continuous valued endpoint). These represent two relatively simple datasets where it is expected that rules will be best suited to modeling the 6-bit multiplexer, while GP trees will be best suited to modeling the additive association with a continuous endpoint.

Our initial evaluation involved running each dataset using a rule population of 1000, and 10,000 learning iterations, with expert knowledge guidance and attribute tracking deactivated for simplicity but all other ExSTraCS run parameters set to their recommended default values as described in [24]. Following training of each dataset, we examined the relative number of GP trees vs. rules as well as the overall prediction accuracy of the rule/tree population.

4.3   Preliminary Results

Initial examination of the rule population following training on the 6-bit and 11-bit multiplexer problems, suggested that ideal rules were being discovered that captured the complex heterogeneous associations in these datasets. Figure 4.4 illustrates LCS rules and GP trees as they exist together on the common pareto-front inspired fitness landscape. Note that ideal rules (in blue) were identified with a perfect accuracy and a normalized coverage of 4 which is expected for the 6-bit multiplexer problem. Alternatively we observe that GP trees as individuals are unable to achieve the same level of accuracy as individual models, however they inherently correctly cover a larger number of individuals in the datasets, since any GP tree 'matches' (i.e. can be applied to make a prediction for) every instance in the training data.

Differently, as illustrated in Fig. 4.5, when run on the continuous endpoint dataset using the same run configuration we observe that a GP tree emerged that quite nearly identified the simple additive underlying association, while a large number of accurate, but low-coverage rules were need to piece-wise identify the same pattern. However, this prototype implementation is still flawed given that the objective metrics of 'Useful Accuracy' and 'Normalized Coverage' were designed exclusively for rules [16, 25]. Briefly, 'Useful Accuracy' is the classification accuracy of a rule beyond what is expected for the given class by random chance, while 'Normalized Coverage' is the number of instances in the datasets that were correctly classified by the rule beyond the number expected by random chance.

This flaw can be visualized in Figs. 4.4 and 4.5 by the parametric curve pattern of the GP model red dots, which would be expected to instead fall on a straight line given that normalized coverage is a linear function of accuracy. This preliminary analysis has revealed that new strategies or uniquely adapted metrics will be required for GP trees in the context of LCS rules and continuous endpoints.

The next task will be to finish adapting this prototype system and run respective comprehensive analyses across a much wider simulation study. Additionally we will update ExSTraCS to output GP trees along side rules for inspection and interpretation in the context of an LCS rule population.

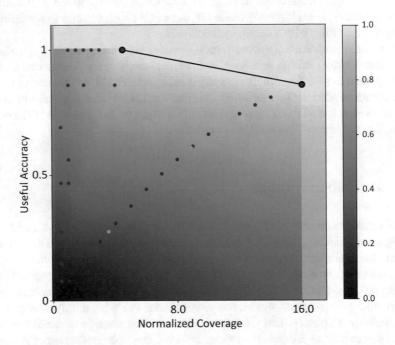

Fig. 4.4 Pareto-front inspired multi-objective entity fitness landscape for the 6-bit multiplexer problem. The colored gradient indicates the respective fitness of entities as a function of the distance from the entity front (defined by the black dots and line). Blue dots represent LCS rules that have trained on the entire dataset, while green dots represent young/inexperienced rules that have not. Red dots represent GP trees that have trained on the entire dataset, while orange dots represent GP trees that have not

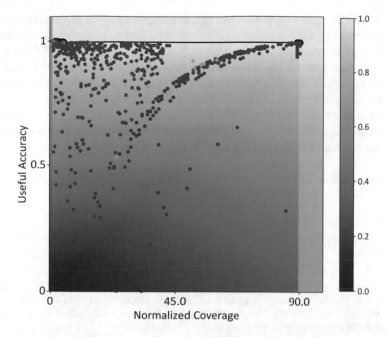

Fig. 4.5 Pareto-front inspired multi-objective entity fitness landscape for the additive continuous endpoint dataset. The colored gradient indicates the respective fitness of entities as a function of the distance from the entity front (defined by the black dots and line). Blue dots represent LCS rules that have trained on the entire dataset, while green dots represent young/inexperienced rules that have not. Red dots represent GP trees that have trained on the entire dataset, while orange dots represent GP trees that have not

4.4   Conclusions and Ongoing Work

In this study we propose a methodology for integrating the respective advantages of evolutionary rule-based machine learning with those of genetic programming. Specifically, we sought to integrate the ability of LCSs to model heterogeneous patterns of association, and complex classification problems, with the ability of GP to compactly model simpler non-heterogeneous patterns, as well as regression problems in a relatively interpretable manner. Our preliminary results support our hypothesis, that by co-evolving both representations, the underlying problem will drive the system to preferentially use the preferable representation. However these initial results were gathered on a prototype implementation and over two small and simple target test problems. There are many alternative implementation considerations that might lead to improved performance. Thus far we can conclude that this proposed approach holds promise and is worth further evaluation and methodological development.

Ongoing work will focus most immediately on adapting this rule/tree implementation further to the challenges of continuous endpoint dataset analysis including a much more extensive evaluation, testing a much wider variety of simulated datasets with different underlying patterns of association. Rigorous statistical analyses and comparisons between ExSTraCS with and without this proposed GP integration over this spectrum of simulated data should also be completed. Further, the composition of the resulting rule populations should be examined to assess solution interpretability with and without GP under different problem scenarios.

Lastly, we expect that it will be beneficial to adapt strategies to utilize expert knowledge and attribute tracking [24] to guide GP tree initialization and evolution in a manner similar to that of LCS rules in our ExSTraCS algorithm.

Acknowledgements This work was supported by NIH grants AI11679, LM009012, AI116794, DK112217, ES013508, and LM010098.

References

1. M. Ahluwalia and L. Bull. A genetic programming-based classifier system. In *Proceedings of the genetic and evolutionary computation conference*, volume 1, pages 11–18, 1999.
2. J. Bacardit, E. K. Burke, and N. Krasnogor. Improving the scalability of rule-based evolutionary learning. *Memetic Computing*, 1(1):55–67, 2009.
3. E. Bernadó-Mansilla and J. M. Garrell-Guiu. Accuracy-based learning classifier systems: models, analysis and applications to classification tasks. *Evolutionary Computation*, 11(3):209–238, 2003.
4. A. Bonarini. An introduction to learning fuzzy classifier systems. In *Learning Classifier Systems*, pages 83–104. Springer, 2000.
5. M. V. Butz and O. Herbort. Context-dependent predictions and cognitive arm control with xcsf. In *Proceedings of the 10th annual conference on Genetic and evolutionary computation*, pages 1357–1364. ACM, 2008.
6. P. G. Espejo, S. Ventura, and F. Herrera. A survey on the application of genetic programming to classification. *IEEE Transactions on Systems, Man and Cybernetics, Part C: Applications and Reviews*, 40(2):121–144, 2010.
7. M. Iqbal, W. N. Browne, and M. Zhang. Xcsr with computed continuous action. In *AI 2012: Advances in Artificial Intelligence*, pages 350–361. Springer, 2012.
8. M. Iqbal, W. N. Browne, and M. Zhang. Evolving optimum populations with xcs classifier systems. *Soft Computing*, 17(3):503–518, 2013.
9. J. R. Koza. Genetic programming ii: Automatic discovery of reusable subprograms. *Cambridge, MA, USA*, 1994.
10. P. L. Lanzi. Extending the representation of classifier conditions part i: from binary to messy coding. In *Proceedings of the 1st Annual Conference on Genetic and Evolutionary Computation-Volume 1*, pages 337–344. Morgan Kaufmann Publishers Inc., 1999.
11. P. L. Lanzi and D. Loiacono. Classifier systems that compute action mappings. In *Proceedings of the 9th annual conference on Genetic and evolutionary computation*, pages 1822–1829. ACM, 2007.
12. A. Orriols-Puig, J. Casillas, and E. Bernadó-Mansilla. Fuzzy-ucs: a michigan-style learning fuzzy-classifier system for supervised learning. *Evolutionary Computation, IEEE Transactions on*, 13(2):260–283, 2009.
13. P. Tufts. Dynamic classifiers: genetic programming and classifier systems. In *Proceedings of the Genetic Programming. Papers from the 1995 AAAI Fall Symposium*, pages 114–119, 1995.

14. R. Urbanowicz, A. Granizo-Mackenzie, and J. Moore. Instance-linked attribute tracking and feedback for michigan-style supervised learning classifier systems. In *Proceedings of the fourteenth international conference on Genetic and evolutionary computation conference*, pages 927–934. ACM, 2012.
15. R. Urbanowicz and J. Moore. The application of michigan-style learning classifier systems to address genetic heterogeneity and epistasis in association studies. In *Proceedings of the 12th annual conference on Genetic and evolutionary computation*, pages 195–202. ACM, 2010.
16. R. Urbanowicz and J. Moore. Retooling fitness for noisy problems in a supervised michigan-style learning classifier system. In *Proceedings of the 2015 Annual Conference on Genetic and Evolutionary Computation*, pages 591–598. ACM, 2015.
17. R. Urbanowicz, N. Ramanand, and J. Moore. Continuous endpoint data mining with exstracs: A supervised learning classifier system. In *Proceedings of the Companion Publication of the 2015 Annual Conference on Genetic and Evolutionary Computation*, pages 1029–1036. ACM, 2015.
18. R. J. Urbanowicz, A. S. Andrew, M. R. Karagas, and J. H. Moore. Role of genetic heterogeneity and epistasis in bladder cancer susceptibility and outcome: a learning classifier system approach. *Journal of the American Medical Informatics Association*, 20(4):603–612, 2013.
19. R. J. Urbanowicz, G. Bertasius, and J. H. Moore. An extended michigan-style learning classifier system for flexible supervised learning, classification, and data mining. In *Parallel Problem Solving from Nature–PPSN XIII*, pages 211–221. Springer, 2014.
20. R. J. Urbanowicz and W. N. Browne. *Introduction to Learning Classifier Systems*. Springer, 2017.
21. R. J. Urbanowicz, D. Granizo-Mackenzie, and J. H. Moore. Using expert knowledge to guide covering and mutation in a michigan style learning classifier system to detect epistasis and heterogeneity. In *Parallel Problem Solving from Nature-PPSN XII*, pages 266–275. Springer, 2012.
22. R. J. Urbanowicz and J. H. Moore. Learning classifier systems: a complete introduction, review, and roadmap. *Journal of Artificial Evolution and Applications*, 2009.
23. R. J. Urbanowicz and J. H. Moore. The application of pittsburgh-style learning classifier systems to address genetic heterogeneity and epistasis in association studies. In *International Conference on Parallel Problem Solving from Nature*, pages 404–413. Springer, 2010.
24. R. J. Urbanowicz and J. H. Moore. Exstracs 2.0: description and evaluation of a scalable learning classifier system. *Evolutionary Intelligence*, 8(2–3): 89–116, 2015.
25. R. J. Urbanowicz, R. S. Olson, and J. H. Moore. Pareto inspired multi-objective rule fitness for noise-adaptive rule-based machine learning. In *International Conference on Parallel Problem Solving from Nature*, pages 514–524. Springer, 2016.
26. M. Valenzuela-Rendón. The fuzzy classifier system: A classifier system for continuously varying variables. In *Proceedings of the Fourth International Conference on Genetic Algorithms (Morgan Kauffman)*, 1991.
27. S. W. Wilson. Classifier fitness based on accuracy. *Evolutionary computation*, 3(2):149–175, 1995.
28. S. W. Wilson. Get real! xcs with continuous-valued inputs. In *Learning Classifier Systems*, pages 209–219. Springer, 2000.
29. S. W. Wilson. Function approximation with a classifier system. In *Proc. 3rd Genetic and Evolutionary Computation Conf.(GECCO'01)*, pages 974–981. Citeseer, 2001.
30. S. W. Wilson. Classifiers that approximate functions. *Natural Computing*, 1(2–3):211–234, 2002.
31. S. W. Wilson. Three architectures for continuous action. In *Learning Classifier Systems*, pages 239–257. Springer, 2007.

Chapter 5
Applying Ecological Principles to Genetic Programming

Emily Dolson, Wolfgang Banzhaf, and Charles Ofria

Abstract In natural ecologies, niches are created, altered, or destroyed, driving populations to continually change and produce novel features. Here, we explore an approach to guiding evolution via the power of niches: ecologically-mediated hints. The original exploration of ecologically-mediated hints occurred in Eco-EA, an algorithm in which an experimenter provides a primary fitness function for a tough problem that they are trying to solve, as well as "hints" that are associated with limited resources. We hypothesize that other evolutionary algorithms that create niches, such as lexicase selection, can be provided hints in a similar way. Here, we use a toy problem to investigate the expected benefits of using this approach to solve more challenging problems. Of course, since humans are notoriously bad at choosing fitness functions, user-provided advice may be misleading. Thus, we also explore the impact of misleading hints. As expected, we find that informative hints facilitate solving the problem. However, the mechanism of niche-creation (Eco-EA vs. lexicase selection) dramatically impacts the algorithm's robustness to misleading hints.

5.1 Introduction

Natural evolution produces effective solutions to complex problems, often well beyond the ability of human engineers to duplicate. If we are to harness these natural evolutionary dynamics, we must understand the full depth of how they function

E. Dolson (✉) · C. Ofria
BEACON Center for the Study of Evolution in Action and Department of Computer Science and Ecology, Evolutionary Biology, and Behavior Program, Michigan State University, East Lansing, MI, USA
e-mail: dolsonem@msu.edu; ofria@msu.edu

W. Banzhaf
BEACON Center for the Study of Evolution in Action and Department of Computer Science, Michigan State University, East Lansing, MI, USA
e-mail: banzhafw@msu.edu

© Springer International Publishing AG, part of Springer Nature 2018
W. Banzhaf et al. (eds.), *Genetic Programming Theory and Practice XV*,
Genetic and Evolutionary Computation, https://doi.org/10.1007/978-3-319-90512-9_5

and why they are so effective. In this paper, we explore how ecological factors promote more open-ended evolutionary systems that have a greater potential to produce complex, dynamic, and practical solutions to targeted problems. First we discuss why we believe that this approach can help solve difficult AI problems, then we review the current scientific understanding of ecological dynamics of interest. Next, we present experiments that we have performed using Eco-EA and lexicase selection to test the potential of ecologically-mediated hints, before discussing the implications of these results, and finally laying out the next steps we plan to take in this line of research.

5.1.1 Motivation

Many problems in machine learning center around creating artificial intelligence systems that can resolve challenging problems that are traditionally solved by humans. Often, these programs are written with the goal of mimicking the strategies employed by skilled humans. When human strategies can be clearly articulated as a set of well-defined rules, the process of writing the AI can be straight-forward. However, humans tend to rely on intuition when solving many types of problems. While human intuition is often effective, it is challenging to abstract into an algorithm that an AI can follow without missing important nuances. Attempts to do so often produce rigid AIs that fail to appropriately adapt to situations that are subtly different than those that were originally expected. Such issues are exacerbated for problems where early decisions determine what scenarios the AI will later encounter. For example, in many board games, minor mistakes early on can drastically reduce success and alter what options are available for the middle and end of the game.

This property—whereby early decisions shape what strategies are possible later—can pose even more substantial obstacles to evolving AIs. If a good strategy is critical early on, an AI without one would consistently lose. However if a good early strategy is not sufficient for overall success, the selection pressure in favor of it will be weak (at best) relative to its importance. Specifically, in any problem where many tasks must all be performed reasonably well for fitness to be non-zero, it is nearly impossible for evolution to get enough initial traction to be successful. Taken, from another perspective, the fitness landscape for such a problem will be almost entirely flat, with a small rugged region featuring steep peaks that are challenging to even find, let alone navigate. Previous attempts to make this landscape more easily navigable by giving the evolving AIs "hints" based on strategies successfully employed by humans have generally been ineffective due to the aforementioned difficulties with abstracting human intuition.

We hypothesize that we can apply ecological dynamics to evolve models of human decision making and reliably create human-competitive AIs. Here, we present our concepts in terms of playing games, as this is an intuitive set of problems to think about, but the techniques we propose should generalize to other categories of problems. The key is to have evolution be responsible for the process of abstracting general strategy from human intuition. We believe that we can achieve this goal by supplying data on human decisions from a wide variety of scenarios and selecting for AIs that are capable of *predicting* the move that a given human made in any given situation. This approach will allow selection to operate evenly across early-, mid-, and late-game strategy, removing the issue of temporally compounded mistakes. Furthermore, it eliminates the need for humans to be able to codify their strategies into a concrete rule set.

Selecting for AIs that can predict human moves has a second, more fundamental benefit. Often, problems that are too complex for evolutionary computation to solve in isolation can be solved if there is also a fitness benefit for solving simpler, related problems. These simpler problems, often referred to as "building blocks", [16] cause genomes within the population to accumulate information that is relatively easy to repurpose into solving the actual problem [16], a technique that has proven to be effective in evolutionary computation [1]. We hypothesize that AI's evolving to predict a human's choices will often do so by recreating the underlying building blocks that comprise that human's strategy—even if the human does not recognize it themself.

An obvious problem with this approach is that some humans may be poor at a given problem. For example, predicting their moves in a game may be impossible because they are effectively random, or worse employing actively poor strategy that will lead the evolving population down a maladaptive path. Fortunately, Goings and Ofria previously developed an evolutionary algorithm that is resistant to bad advice: Eco-EA [5–7]. Eco-EA creates limited resources associated with tasks that are expected to be valuable on the way to a high-quality solution. This approach proved effective on example problems, in part because human intuition could be used to select the rewarded tasks and the algorithm would associate limited resources with them. As long as no organisms are performing the rewarded traits, the resources build up in the population (to a limit) and make the associated tasks more valuable. Once the resources are in use, however, their abundance dwindles until they supply fitness to only a small fraction of the population. Resources associated with useful sub-tasks should produce evolutionary building-blocks that get incorporated as parts of a more complex strategy and used in the overall solution. Unhelpful or misleading resources, however, should be used by only a small fraction of the population, resulting in a trivial slowdown as compared to running the algorithm without the resources present.

For example, if we wanted to evolve a controller for a robot that would perform search-and-rescue, we would never get there if we started with random controllers and only rewarded robots that successfully rescued people. We could, however, use ecologically-mediated hints by adding a set of limited resources that each rewarded some component, such as (1) moving to a target location, (2) exploring, (3) fully scanning an area, (4) identifying dangers, (5) navigating around obstacles, (6) identifying trapped people, (7) moving toward trapped people, (8) freeing trapped people, (9) moving with people, and (10) finding your way back to safety. With all of these helpers as building blocks, it is easy to imagine that a controller would eventually evolve that could occasionally find and return people safely, which would allow it to start getting rewarded by the primary (unlimited) fitness function. This approach uses ecologically-mediated hints to push the population to solve the problem from multiple directions at once; before the whole problem is solved, some individuals might be able to find people, but then ignore them. Some will explore and then find their own way back. Others might be able to free people they find, but not know what to do with them next. Some evolutionary trajectories might be more likely than others to continue to evolve toward the final goal. Of course, some of the "hints" might be associated with counter-productive tasks. If there were a limited resource associated with avoiding all dangers (because a misguided programmer thought this would be helpful), the result might be to make it less likely for the robot to find trapped people. With ecologically-mediated hints, however, only a small portion of the population (a single niche) would be rewarded for this ability, so others would still plow ahead and complete the mission.

The inspiration for this approach comes from the study of eco-evolutionary dynamics [11]. Ecology and Evolution are considered sister fields of study within Biology, often separated by the misconception that their dynamics occur on two different timescales. However, the lines between these disciplines have blurred as it has been acknowledged that ecological and evolutionary dynamics strongly influence one another [21]. Practitioners of evolutionary computation have recognized this relationship, but typically use ecological effects only for preserving diversity to prevent premature convergence on a sub-optimal solution and produce a wider range of solutions to choose among. In Artificial Life systems, however, ecology has also been linked to promoting open-ended evolution in the form of increased complexity, novelty, cooperation, distributed problem solving, and intelligence [24]. As such, we believe that richer ecological dynamics are a source of substantial untapped potential for evolutionary computation.

A key component of this approach is that in an ecology, niches drive the types of diverse solutions that appear. If an organism is the first to occupy a new niche, it must have some traits associated with that niche, but is otherwise free from direct competition, allowing it to sustain substantial loss of unrelated traits. This dynamic allows for many different pathways to coexist in a population, any of which can be followed to solve the high-level problem, with the most successful strategies dominating multiple niches. In essence, these co-existing niches facilitate the creation of a variety of building blocks leading to successively more complex strategies.

Another benefit to an eco-evolutionary approach is that ecological communities are, by definition, at least somewhat diverse. While promoting diversity in evolutionary computation has long been recognized as critical to avoiding the problem of premature convergence, most existing mechanisms to promote diversity in evolutionary computation select for solutions that are distinct from each other, regardless of other qualities [8]. In natural systems, however, diversity arises due to organisms filling niches, each requiring specific phenotypic traits for success. Thus there is pressure for a diversity of functional traits. Furthermore, new niches are continuously created in nature as organisms interact with each other and modify both their physical and social environment. In problem domains such as playing complex games, this diversity of solutions becomes even more important, as there is no single, deterministic, best strategy. Instead, there are various strategies that are effective in different situations and against different opponents, just as there would be in an ecological community.

5.1.2 *Ecological Approaches in Evolutionary Algorithms*

A number of highly effective evolutionary-computation techniques owe their success to ecological dynamics. First, there is lexicase selection [22], a different approach to creating niches associated with various sub-problems, which has proven extraordinarily successful in genetic programming [9, 10]. In lexicase selection, a large number of test cases are used as criteria for evaluation. Each time an organism has to be selected to propagate into the next generation, the test cases are applied in a random order, filtering out all but the most fit candidates each time. Once a single candidate remains (or all of the test cases have been applied and a random individual is chosen from the final set) it is replicated into the next generation and the process is repeated. The fact the ordering of these test cases continually changes means that solutions successful at different subsets of tasks are all able to co-exist. Although lexicase selection has traditionally been used with test cases, there should be no reason it cannot be used with multiple fitness functions instead to provide ecologically-mediated hints. In this way, any hint that could be given to a system like Eco-EA could just as easily be provided to lexicase selection as well.

From an ecological perspective, lexicase selection creates pressure for the population to diversify into different niches that are based on building blocks to a more complex problem. In order for an organism to be evolutionarily successful, it must be among the very best at at least one of the test cases/fitness functions. In most cases, it must be among the best at many of them (probably a related group). Thus, organisms compete with others within the niches created by each test case/fitness function. Eco-EA also promotes intra-niche competition, but in a subtly different way. Limited resources are used up as organisms receive fitness bonuses, meaning that they compete with other organisms benefiting from the same bonus. Thus, those with higher fitness will competitively exclude less fit organisms that depend too much on the same resources [2].

This distinction between competition in lexicase selection and in systems with limited resources has a few implications. First, limited resources allow for generalists that do not excel at any tasks but are decent at many. It is unclear whether this flexibility opens up additional pathways through the fitness landscape. Second, lexicase selection always exerts selective pressure to improve on all test cases/fitness functions, whereas limited resources create little incentive to expand into a niche that is already fully occupied. This dynamic may make lexicase selection less robust to receiving bad advice about how to solve a problem than other mechanisms for providing ecologically-mediated hints. Third, lexicase selection automatically incentivizes organisms that excel at uncommon combinations of test cases/fitness functions. Eco-EA could, however, be adjusted to mimic this property by adding additional resources. Ideally, we will be able to find a hybrid approach to providing ecologically-mediated hints that combines Eco-EA with lexicase selection to achieve the best of both worlds.

MAP-Elites is another successful algorithm that leverages ecological dynamics and user input [19]. In MAP-Elites, the user chooses axes along which they believe variation will be important. MAP-Elites then breaks the hyperspace defined by these axes into discrete regions. Each region can be occupied by at most one solution. When a new candidate solution is created, it is assessed on each axis and the corresponding bin is located. If the new candidate solution has a higher fitness than the previous occupant of that bin, it replaces that occupant. Otherwise, it is tossed out. Evolution proceeds by choosing occupants of various bins in the search space and mutating them to create new candidate solutions.

The sub-division of axes for MAP-Elites explicitly partitions the search space into niches in a way that has clear parallels to choosing tests to provide to lexicase selection or hints to associate with limited resources. The lack of directionality in MAP-Elites, however, makes it distinct from lexicase selection. At face value, it might seem to make it different from Eco-EA as well. Certainly this intuition is true to some extent—hints provided as limited resources do have directionality to them. However, the foundation of limited resources is negative frequency-dependence which, in addition to reducing the risk of the entire population being led astray, has the important effect of promoting diversity along the axis of the hint. This diversity goes beyond merely allowing a portion of the population to drift; niche partitioning should force the population to spread out along the axis in question to produce meaningful diversity that may be helpful in solving the overall problem.

There are a wide variety of other ecologically-inspired strategies that have proven to be effective for maintaining diversity in evolutionary algorithms. These largely fall into four categories: niching/speciation (e.g. [8, 23]), parent selection (e.g. [4, 17]), dividing the population into subpopulations (e.g. [13, 14]), and adjusting the objective function to favor diversity and/or novelty (e.g. [18]). Of these approaches, only niching/speciation consistently promotes stable coexistence of different types of strategy. Even among niching/speciation strategies, most emphasize a diversity of phenotypes rather than the diversity of evolutionary building blocks that Eco-EA and Lexicase Selection promote. While solutions built on different building blocks may often exhibit different phenotypes, they may also arrive at similar phenotypes despite taking very different paths through the fitness landscape. These different paths will likely result in underlying differences in genetic architecture that may influence which behaviors are easy for a lineage to evolve next. As such, we suspect that diversity of building blocks will promote greater evolutionary potential than other kinds of population diversity.

5.1.3 Limited Resources and Eco-EA

The original formulation of limited resources that led to ecologically-mediated hints arose in work by Cooper and Ofria where they demonstrated that limited resources are sufficient to evolve stable branching of different ecotypes [3]. The resulting ecological community was incredibly simple, with competition being the only form of interaction between organisms. However, we hypothesized that these simple ecologies (with meaningful differences between niches) were better positioned to solve complex problems than populations in which diversity was promoted by other means.

Goings and Ofria initially tested a more applied form of limited resources in Eco-EA, using a toy bitstring matching problem, with the goal of maintaining a population of diverse solutions [7]. In this experiment, resources (either limited or unlimited) were associated with different bitstrings that could be matched to varying extents by members of the population. When resources were unlimited, the population converged on matching a single one of these strings. Limiting the resources, however, produced consistent subpopulations specialized on each string. The negative frequency dependent selection imposed by the limited resources caused the different bitstrings to stably coexist within the population. These results held when the task was to match a more general pattern of bits, rather than an exact string.

Following this initial conceptual test, Goings et al. applied Eco-EA to a complex real-world problem: generating models for the behavior of sensor nodes in a flood warning network [6]. Not only did Eco-EA successfully evolve a diversity of models that satisfied the constraints of the problem, but the models evolved by Eco-EA were better than a comparison algorithm at continuing to diversify when transferred to an environment without limited resources. These results demonstrate the strength of Eco-EA at evolving a diversity of solutions to complex problems.

Eco-EA is just one example of the benefits that making more thoughtful and intentional use of ecology can provide. Here, we define ecological dynamics as interactions between members of the population that affect fitness. We propose that such dynamics provide a variety of benefits, some of which have begun to be put in to practice and some of which have not. The most obvious of these effects is the benefit to diversity alluded to above.

5.1.4 Complexifying Environments

Thus far, the ecological communities that we have discussed have only contained competitive interactions between organisms. In nature, however, there are many other kinds of interaction, such as mutualism, parasitism, and predation. The network of interactions within a community grows increasingly complex over evolutionary time, as the evolution of new species creates new niches for other species to evolve into. This gradual complexification has two implications that may be important for evolutionary computation: (1) The gradual increase in complexity of the biotic environment facilitates the continual production of more complex niches and thus the evolution of more complex traits, and (2) Ecological communities as a whole can often perform functions that no single species could perform alone.

The niches inhabited by various members of an ecological community usually bear commonalities. While they are not identical, the ability to occupy one niche is often a building block for the ability to occupy a different one (a clear example of this effect in nature is metabolic pathways for metabolizing various resources). When new niches appear, they are closely enough related to existing niches that it's plausible that existing populations will evolve to inhabit them. Without the gradual feedback loop of increasing complexity that eco-evolutionary feedbacks enable, evolving the ability to exploit newly established niches would be incredibly improbable.

Community-level functions represent a different mechanism for solving complex problems. For example, a forest, collectively, is able to store solar energy in sugar molecules, fix nitrogen, take in various nutrients from the soil, and use these nutrients and energy to power mobile, decision-making agents. Of course, these functions are each performed by different species in the community. While it may be possible to evolve a single species that carries out all of these tasks, it is worth considering the possibility that it is easier to evolve a community that carries them out collectively. The fact that such communities seem to be so common in nature is certainly suggestive of this possibility. Such an approach would have a variety of potential benefits. For example, it would promote more functional modularity, a trait that is believed to be critical for evolutionary potential. Moreover, the closely interdependent species found in a complex ecological community are a likely precursor to egalitarian major transitions, which are believed by many to be a critical step towards evolving complex species. Lastly, as ecological communities can effectively be thought of as a set of subroutines running in parallel, they lend themselves easily to evolving parallel programs, somewhat akin to Holland's bucket brigade algorithm [12] and more recent advances in learning classifier systems.

Incorporating such complex interactions into evolutionary algorithms will be challenging, and is far beyond the scope of the current paper. However, these factors are yet another reason that we believe that ecological dynamics have the potential to be a powerful force in evolutionary computation in the long run.

5.2 Methods

We expect that ecologically-mediated hints provided via Eco-EA and lexicase selection will display similar set of strengths. Both maintain a diverse population with respect to the hints that are provided. However, as discussed above, we expect the use of limited resources to be robust to hints that turn out to be counter-productive, while lexicase selection will have difficulty escaping uncontested bad advice. To assess the accuracy of these hypotheses, we evaluate them in the context of a proof-of-concept problem where we can easily manipulate the quantity and quality of the hints.

5.2.1 10-Dimensional Box Problem

The search space for our proof-of-concept problem is a 10-dimensional box. All sides of the box have a length of 1. Candidate solutions in the population are sequences of 10 floating point numbers between 0 and 1, representing a point within the box. The goal is to find the origin (i.e. the sequence [0, 0, 0, 0, 0, 0, 0, 0, 0, 0]). This problem could be trivially solved by using the inverse Euclidean distance between a point within the box and the origin as the fitness function. To simulate a more challenging problem, we use the following fitness function:

$$
fitness = \begin{cases} 0.01 & \sqrt{\sum_{i=1}^{10} x_i^2} > 0.1 \\[2em] \dfrac{1}{\sqrt{\sum_{i=1}^{10} x_i^2}} & \sqrt{\sum_{i=1}^{10} x_i^2} \le 0.1 \end{cases} \tag{5.1}
$$

In this function, inverse Euclidean distance from the origin is the fitness only when Euclidean distance from the origin is less than 0.1 (which, given the high dimensionality of the space, represents 2.5×10^{-13} of the possible positions). For all points in the box that are farther away from the origin, fitness is 0.01. We used 0.01 rather than 0 as the base fitness to ensure the Eco-EA's fitness multipliers would have an effect. The cut-off of 0.1 was chosen such that it is unlikely for evolution to solve the problem without hints. Using this fitness function creates a quintessential "needle in a haystack" problem, where the fitness landscape is flat except for one incredibly tall and thin peak. Such problems are generally considered to be among the most challenging for evolutionary computation to solve, as they do not allow for incremental improvement.

To make this problem more possible for evolution to solve, we can provide hints about the optimal value for each dimension. Since the goal is to minimize all dimensions, a good (i.e. informative) hint would be to minimize an individual dimension. Conversely, maximizing an individual dimension would be a bad (i.e. misleading) hint, leading a population away from the goal. We provided various combinations of good and bad hints to Eco-EA, lexicase selection, and to standard tournament selection. As an additional control, we ran an equivalent number of trials using unaltered tournament selection.

5.2.2 Eco-EA Implementation

The crux of the idea behind Eco-EA is that it must provide hints about how to solve a problem and incentivize following them in a manner that leads to negative frequency-dependence. There are a variety of ways to implement this concept, and there are likely various trade-offs amongst them that are worthy of a systematic study. For the purposes of this chapter, we have used an implementation as similar as possible to the original Eco-EA implementation [7].

Each hint is associated with a resource. That resource flows into the environment at some rate, I, and flows out at some rate, O. For these experiments, the inflow rate, I, was 100, and the outflow rate, O, was 0.01. Thus, 100 units of each resource entered the environment over the course of each generation, and 1% of the total quantity of each resource exited the environment at the end of each generation. Resources entered the environment at a constant rate as fitnesses were evaluated to minimize stochastic effects from the order in which solutions were evaluated.

There are a few parameters that are necessary for determining how resources are consumed and how they impact fitness. First, there is C_f, the consumption fraction. Just as no organism in an ecosystem in nature is capable of consuming all of a resource in that environment, no solution in Eco-EA is. C_f specifies what fraction of the total quantity of resource any individual solution consumes. Next, there is m, the maximum amount of resource an individual is capable of consuming at any one time. For all experiments, we use $C_f = 0.0025$ and $m = 5$ to maintain consistency with previous Eco-EA research [5]. Additionally, there is the c, the cost of attempting to use a hint. In order to create negative frequency dependence, there must be a cost to attempting to use a hint when too many other members of the population are also attempting to use it. In most scenarios, there is an implicit cost to attempting to use a hint, stemming from the trade-offs inherent in choosing to take one action over another. In a problem as simple as the 10-dimensional box problem, however, there is no such implicit cost. Thus, without an explicit cost, there will not be negative frequency-dependence. Since negative-frequency dependence is a core element of Eco-EA, we impose a cost of $c = 1$. Lastly, if there is a cost to attempting to use a hint, then we also must define the range within which a solution is considered to be attempting to use that hint. We call this range the niche width, n, which is set to 0.2 for all of these experiments. Since the maximum score for any hint function is 1, this means that solutions must score a minimum of 0.8 on a hint function in order to potentially get a reward or pay a cost from it.

In order to calculate the fitness impact of a hint, we need two more pieces of information. The first is R, the current amount of the relevant resource present. The second is s, the score on the hint function, which is squared to incentivize even small increases in performance on the hint function. The amount of resource successfully used, A, is calculated with the equation:

$$A = \begin{cases} 0 & s < 1 - n \\ min((s^2 * C_f * R) - c, m) & s \geq 1 - n \end{cases} \quad (5.2)$$

A is subtracted from the current amount of resource in the environment, and is used to update the base fitness of the current organism with the equation:

$$fitness = fitness * 2^A \quad (5.3)$$

Thus, successfully using more than c resource will multiply the solution's fitness by a number greater than one, whereas using less than c resource will multiply the solution's fitness by a number less than one.

These calculations are performed for all hints for all members of the population to determine resource-adjusted fitness for all candidate solutions. Tournament selection with a tournament size of two is then performed on the population based on these fitness values. To ensure that the impact of hints doesn't wash out small fitness gains on the main fitness function, every generation created with Eco-EA also contains a copy of the individual from the previous generation with the highest base fitness.

5.2.3 Lexicase Selection Implementation

Traditionally, lexicase selection is given a large number of test cases. For each iteration of selection, these test cases are placed in a random order. Each member in the population is evaluated on each one in sequence. For each test case, only those solutions that performed best are kept in contention to be selected. When only a single solution is left, it is placed into the next generation. Ties are broken randomly.

We argue that test cases can be thought of as a subset of the broader category of hints about how to solve a problem, and that lexicase selection should generalize to any kind of hint. So, in place of test cases, we use the hints on solving the 10-dimensional box problem described above. For every selection event, we randomly order these hint fitness functions (along with the overall fitness function) and filter the population based on this ordering.

5.2.4 Tournament Selection Implementation

In an iteration of tournament selection, a pre-determined number (two, in these experiments) of individuals are chosen at random from the population. The fittest of them is selected to reproduce. In this paper, we use tournament selection as our control, as it is effectively a less-informed version of the implementation of Eco-EA that we use here.

5.2.5   Configuration Details

For all experiments, we evolved a population of 5000 vectors containing 10 floating-point numbers between 0.0 and 1.0 (inclusive) for 50,000 generations. The next population for each generation was chosen using one of the three selection schemes being compared: lexicase selection, Eco-EA, or tournament selection. To avoid giving Eco-EA an unfair advantage due to its use of elitism (in a problem domain where that could only be beneficial), we always preserved a copy of the individual in the population with the highest fitness. For selection schemes requiring a tournament size (tournament selection and Eco-EA), we used a tournament size of two.

We placed the individuals selected by each iteration of the selection scheme into the next generation. Each site in each genome was mutated by adding a value randomly selected from a Gaussian distribution centered at 0 with a standard deviation of 0.05. Subsequently, we recombined each vector with a random other vector, using one-point crossover.

5.2.6   Statistical Methods

To determine the effects of good advice, bad advice, and selection scheme on the probability of solving the problem, we performed a logistic regression. The predictor variables were the number of good hints, the number of bad hints, a boolean indicating whether lexicase selection was used, and a boolean indicating whether Eco-EA was used. Tournament selection, being the control, was the base case to which all other conditions were compared. The regression coefficients for the interactions between good or bad hints and the selection type were used as the test statistic for all statements about the effect of a hint type on a selection scheme.

All statistics were computed using the R statistical computing language, version 3.4.3 [20]. Since some combinations of variables were able to perfectly separate successes from failures, which can bias the results of logistic regression, we used Firth's bias reduction technique, as implemented in the R package brglm [15]. All plots were made using the ggplot2 package [25].

5.2.7   Code Availability

The full source code for the experiments and analysis in this paper is freely available at https://github.com/emilydolson/eco-ea-box. This code makes heavy use of the evolutionary computation modules in the Empirical library, which is available at https://github.com/devosoft/Empirical. All code for this paper is open source and freely available.

5.3 Results and Discussion

Both Eco-EA and lexicase selection make effective use of good advice (see Fig. 5.1). Eco-EA solved the problem consistently when given at least four good hints. Lexicase selection did even better, solving the problem most of the time when given as few as two good hints. Good hints enable both lexicase selection and Eco-EA to significantly outperform tournament selection (logistic regression, $\beta = 1.3334$ for lexicase, $\beta = 1.8363$ for Eco-EA, $p < 0.0001$ for both).

These data strongly support our hypothesis that Eco-EA is essentially unaffected by receiving bad hints; the regression coefficient for the interaction between Eco-EA and bad hints is not significantly different from 0 (logistic regression, $\beta = 0.0633$, $p = 0.66$). Lexicase selection, on the other hand, is harmed dramatically by bad hints (logistic regression, $\beta = -10.6841$, $p < 0.0001$).

Overall, our results illustrate the power of ecological approaches as a vehicle for providing hints to evolutionary algorithms. We have provided a proof-of-concept that this technique can make it possible for evolution to solve problems that would otherwise have been out of reach. Moreover, we have clarified the instances in which two specific ecological approaches, Eco-EA and lexicase selection, are most appropriate; lexicase is best when all of the hints are accurate, whereas Eco-EA is robust in scenarios where there may be some misleading hints.

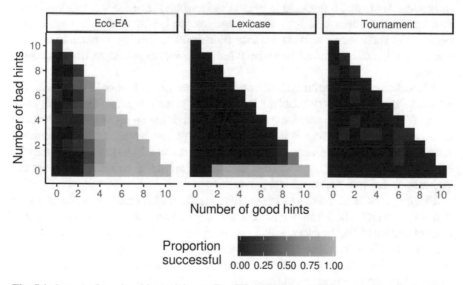

Fig. 5.1 Impact of good and bad advice on Eco-EA and Lexicase. Heat maps for each algorithm show the success rate of that algorithm in the presence of varying quantities of good and bad advice. Note that tournament selection is not actually capable of receiving hints; it is presented here as a control. Each cell in the heat maps represents the proportion of runs (out of 10) that successfully found the optimal solution to the 10-dimensional box problem

5.4   Conclusions and Future Work

Thus far, we have tested ecologically-mediated hints on only simple model problems as a proof-of-concept, but we expect this approach to excel on complex problems where fitness functions do not provide a clear path from random starting conditions to meaningful solutions. Artificial Intelligence is a perfect example of this type of problem. Problem-solving strategies may require many different components to coordinate to formulate plans. Board-game-playing agents are a particularly accessible problem that has these properties: they are complex, while still being experimentally tractable, and are often well-studied. They involve clear measures of success, while allowing for multiple co-existing strategies. Most importantly, they are intuitive to humans who can provide "suggestions" for limited resources to produce building blocks for complex strategies.

The next step will be to more thoroughly explore the parameter space in which providing hints via Eco-EA and lexicase selection is effective. In the process, we hope to gain insight that will allow us to develop an approach that includes the best properties of both. Thus far, we have explored the impact of good and bad hints that are orthogonal to each other. However, most hints in real world problems will not be independent, and we expect that this connection may influence the way Eco-EA and lexicase selection respond to them. Similarly, we have not explored the impact of completely neutral hints, which may be harmful if they are present in excessive quantities. That said, a resource is helpful only when it produces a building block that turns out to be useful for a more complex goal, after which it is no longer needed. As such, many resources would be helpful if they were around just long enough for the associated niche to be filled as a stepping-stone to more complex niches.

If good advice is helpful and other advice is harmless, it should be possible to bootstrap the solving of a problem by generating random hints. Each hint, once used, will remain for a limited amount of time (providing an opportunity to be used as a building-block) before it is removed and replaced by a new hint associated with a new randomly-determined behavior. These random behaviors will usually be harmful and thus ignored, but any that turn out to be helpful building blocks should be incorporated into the successful players.

Finally, we will use human problem-solving patterns to create niches for evolving "crowd-sourced" AIs. While we believe transient resources will have some utility, most of those building blocks are likely to be useless, merely using up computational time. A benefit of working with board games, however, is that we can collect a wealth of information about how human players make decisions across a variety of situations. Rather than reward building blocks that we identify from those logs, we can instead create limited resources that reward AI players for consistently predicting the next move made by a human player. In other words, we do not need to understand why a human player made a decision in order to reward an AI for following the same type of strategy. If that player plays well, the building blocks produced to mimic them should be generally useful, even for some other strategy types. Ultimately, the biggest rewards will still come from winning games, so mimicking poor players should have a minimal negative impact (as is usually the case with Eco-EA).

References

1. Bongard, J. C. and Hornby, G. S. (2010). Guarding Against Premature Convergence While Accelerating Evolutionary Search. In *Proceedings of the 12th Annual Conference on Genetic and Evolutionary Computation*, GECCO '10, pages 111–118, New York, NY, USA. ACM.
2. Chesson, P. (2000). Mechanisms of Maintenance of Species Diversity. *Annual Review of Ecology and Systematics* 31:343–366.
3. Cooper, T. F. and Ofria, C. (2002). Evolution of stable ecosystems in populations of digital organisms. In *Artificial Life VIII: Proceedings of the Eighth International Conference on Artificial life*, pages 227–232.
4. De Jong, K. A. (1975). Analysis of the behavior of a class of genetic adaptive systems.
5. Goings, S. (2010). Natural niching: Applying ecological principles to evolutionary computation. *Dissertation*, Michigan State University.
6. Goings, S., Goldsby, H. J., Cheng, B. H., and Ofria, C. (2012). An ecology-based evolutionary algorithm to evolve solutions to complex problems. *Artificial Life*, 13:171–177.
7. Goings, S. and Ofria, C. (2009). Ecological approaches to diversity maintenance in evolutionary algorithms. In *IEEE Symposium on Artificial Life, 2009. ALife '09*, pages 124–130.
8. Goldberg, D. E. and Richardson, J. (1987). Genetic algorithms with sharing for multimodal function optimization. In *Genetic algorithms and their applications: Proceedings of the Second International Conference on Genetic Algorithms*, pages 41–49. Hillsdale, NJ: Lawrence Erlbaum.
9. Helmuth, T. and Spector, L. (2015). General Program Synthesis Benchmark Suite. *Proceedings of the 2015 Annual Conference on Genetic and Evolutionary Computation*, GECCO '15, pages 1039–1046, New York, NY, USA, ACM.
10. Helmuth, T., Spector, L., and Matheson, J. (2015). Solving Uncompromising Problems With Lexicase Selection. *IEEE Transactions on Evolutionary Computation*, 19(5):630–643.
11. Hendry, A. P. (2016). *Eco-evolutionary Dynamics*. Princeton University Press.
12. Holland, J. H. (1985). Properties of the bucket brigade. In *Proceedings of an International Conference on Genetic Algorithms*, pages 1–7, Hillsdale, NJ, USA. Lawrence Erlbaum Assoc.
13. Hornby, G. S. (2006). ALPS: The Age-layered Population Structure for Reducing the Problem of Premature Convergence. In *Proceedings of the 8th Annual Conference on Genetic and Evolutionary Computation*, GECCO '06, pages 815–822, New York, NY, USA. ACM.
14. Hu, J., Goodman, E., Seo, K., Fan, Z., and Rosenberg, R. (2005). The Hierarchical Fair Competition (HFC) Framework for Sustainable Evolutionary Algorithms. *Evolutionary Computation*, 13(2):241–277.
15. Kosmidis, I. (2017). brglm: Bias Reduction in Binary-Response Generalized Linear Models, version 0.6.1 http://www.ucl.ac.uk/~ucakiko/software.html
16. Lenski, R. E., Ofria, C., Pennock, R. T., and Adami, C. (2003). The evolutionary origin of complex features. *Nature*, 423(6936):139–144.
17. Mahfoud, S. W. (1992). Crowding and preselection revisited. *Urbana*, 51:61801.
18. Mouret, J.-B. and Doncieux, S. (2009). Using Behavioral Exploration Objectives to Solve Deceptive Problems in Neuro-evolution. In *Proceedings of the 11th Annual Conference on Genetic and Evolutionary Computation*, GECCO '09, pages 627–634, New York, NY, USA. ACM.
19. Mouret, J.-B. and Clune, J. (2015). Illuminating search spaces by mapping elites. In *arxiv:1504.04909*

20. R Core Team. (2017). R: A Language and Environment for Statistical Computing. https://www.
    R-project.org
21. Schoener, T. W. (2011). The Newest Synthesis: Understanding the Interplay of Evolutionary
    and Ecological Dynamics. *Science*, 331(6016):426–429.
22. Spector, L. (2012). Assessment of problem modality by differential performance of lexicase
    selection in genetic programming: a preliminary report. In *Proceedings of the 14th annual
    conference companion on Genetic and evolutionary computation*, pages 401–408. ACM.
23. Stanley, K. and Miikkulainen, R. (2004). Competitive coevolution through evolutionary
    complexification. *J. Artif. Intell. Res. (JAIR)*, 21:63–100.
24. Taylor, T., Bedau, M., Channon, A., Ackley, D., Banzhaf, W., Beslon, G., Dolson, E., Froese,
    T., Hickinbotham, S., Ikegami, T., McMullin, B., Packard, N., Rasmussen, S., Virgo, N.,
    Agmon, E., Clark, E., McGregor, S., Ofria, C., Ropella, G., Spector, L., Stanley, K. O., Stanton,
    A., Timperley, C., Vostinar, A., and Wiser, M. (2016). Open-Ended Evolution: Perspectives
    from the OEE Workshop in York. *Artificial Life*, 22(3):408–423.
25. Wickham, H. (2009). *ggplot2: Elegant Graphics for Data Analysis*. Springer-Verlag, New
    York.

Chapter 6
Lexicase Selection with Weighted Shuffle

Sarah Anne Troise and Thomas Helmuth

Abstract Semantic-aware methods in genetic programming take into account information about programs' performances across a set of test cases. Lexicase parent selection, a semantic-aware selection, randomly shuffles the list of test cases and places more emphasis on those test cases that randomly appear earlier in the ordering than those that appear later in the ordering. In this work, we explore methods for weighting this shuffling of test cases to give some test cases more influence over selection than others. We design and test a variety of weighted shuffle algorithms and methods for weighting test cases. In experiments on two program synthesis benchmark problems, we find that none of these methods significantly outperform regular lexicase selection. We analyze these results by examining how each method affects population diversity, and find that those methods that perform much worse also have significantly lower diversity.

6.1 Introduction

Many different types of problems typically tackled by genetic programming (GP), including symbolic regression, classification, and program synthesis, require a program that performs well on a set of tests, which we will call *test cases*. On such problems, each program is evaluated on each test, producing an *error vector* that summarizes its performance on the tests. These error vectors typically provide all of the information used to determine which individuals in the population are selected to be parents of the next generation.

In many parent selection methods, such as the pervasive tournament selection, each error vector is aggregated into a single fitness value that represents the

S. A. Troise
Washington and Lee University, Lexington, VA, USA
e-mail: troises19@mail.wlu.edu

T. Helmuth (✉)
Hamilton College, Clinton, NY, USA
e-mail: thelmuth@hamilton.edu

© Springer International Publishing AG, part of Springer Nature 2018
W. Banzhaf et al. (eds.), *Genetic Programming Theory and Practice XV*,
Genetic and Evolutionary Computation, https://doi.org/10.1007/978-3-319-90512-9_6

performance of an individual on the problem. Such methods ignore a wide range of behavioral and semantic information that could potentially be used to more effectively guide search [16, 22]. Recently, researchers have started incorporating this information in their GP systems, such as in the case of geometric-semantic GP [23], behavioral programming [15], and other semantic-aware methods [19].

One recent semantic parent selection method, lexicase selection, has been shown to improve problem-solving performance on a range of problems compared to tournament selection [8, 9] and other semantic-based selection methods [19]. These encouraging results suggest not only that lexicase selection deserves careful analysis of how it contributes to these improves results, but also whether there are modifications that could be made in order to improve its performance further. In this study, we explore variants of lexicase selection in which we modify how it considers the test cases and their order.

An essential part of the lexicase selection algorithm consists of randomly shuffling the test cases. It then considers the test cases in the shuffled ordering, with test cases earlier in the ordering receiving more attention than those later in the ordering. Traditionally, this shuffling has been conducted in a uniform fashion, with each test case having equal probability of appearing at any position [25]. While many people have asked us personally if it would be useful to weight the shuffling so that some test cases are more likely to come earlier in the shuffling than others, to our knowledge this has not been tested in practice.

In this paper we explore the idea of weighting the shuffle of test cases in lexicase selection. One key question, that does not seem to have an obvious theoretical answer, is how should the test cases be weighted? Should easier test cases appear earlier in the ordering, or should harder cases appear earlier? We could imagine it being better for easier test cases to appear earlier, since this may allow evolution to make small steps to improve slowly over time. On the other hand, maybe it would be better to have harder test cases appear earlier, which could reward programs that perform well on test cases on which the rest of the population performs poorly. Since we do not know the best method for weighting shuffle, here we conduct an empirical investigation of a variety of methods, some of which place easier test cases earlier, some of which place harder test cases earlier, and some of which dictate order based on variance.

Our experiments on two program synthesis problems show a surprising result: while some of the weighting methods reduce the performance of lexicase selection, none of them significantly improve performance. To help explain this result, we examine how each method affects population diversity throughout each GP run. We find that many of the methods result in significant reductions in diversity, and none appear to increase diversity compared to regular lexicase selection. Since we believe that diversity maintenance is an important feature of lexicase selection, these results help explain the cases where shuffling methods perform much worse.

In the next section, we give a detailed description of lexicase selection and prior results that use it. In Sect. 6.3, we describe the weighted shuffling algorithms and our methods for weighting test cases. Next, we discuss the design of our experiments on weighted shuffle, and present results from those experiments. We finally give some examples of related parent selection techniques.

6.2 Lexicase Selection

Lexicase selection is defined in terms of *test cases*, i.e. the data points used to evaluate the performance of individuals in the population. While we treat test cases as input/output pairs of the form used in supervised learning, lexicase selection could work in any population-based search technique where individuals are evaluated on multiple metrics. Lexicase parent selection was motivated by the desire of having parent selection treat individual test cases separately, without ever comparing the results of programs on one test case with the results on another [9, 25].

Algorithm 6.1 presents the lexicase selection algorithm. During lexicase selection, we consider one test case at a time, whittling down the population by removing any individual that does not exhibit the very best performance on that case. Once a single individual remains, it is returned. If we iterate through every test case and multiple individuals remain, that means those individuals have identical error vectors, so we return one of them at random. In practice, we actually retain only one random individual per error vector prior to each lexicase selection, since this gives the exact same results and reduces the time required to filter the population at each step.

A key element of the lexicase selection algorithm is that the test cases are shuffled into a different order for selecting each parent. The test cases at the start of the shuffled list have the most impact on selection, since they have potential to filter out the most individuals from the pool. Many times, a test case near the end of the shuffled list will have no bearing on which individual is selected, if the set of candidates is whittled to a single individual before using every test case. In this way,

---

Algorithm 6.1 Lexicase selection *(to select one parent)*

Inputs: *candidates*, the entire population; *cases*, a list of test cases
Shuffle *cases* into a random order
loop
    Set *first* be the first case in *cases*
    Set *best* be the best performance of any individual currently in *candidates* on *first*
    Set *candidates* to be the subset of *candidates* that have exactly *best* performance on *first*
    if $|candidates| = 1$ **then**
        Return the only individual in *candidates*
    end if
    if $|cases| = 1$ **then**
        Return a randomly selected individual from *candidates*
    end if
    Remove the first case from *cases*
end loop

---

lexicase selection often selects *specialist* individuals that perform poorly on some cases as long as they perform very well on the cases at the start of the ordering [4].

Empirical studies have shown lexicase selection to increase and maintain much higher levels of behavioral diversity than tournament selection [5, 6]. These effects on diversity are thought to be a consequence of lexicase selection's emphasis on selecting different specialist individuals. In particular, since lexicase selection uses a different ordering of test cases for each selection, it is able to reward individuals that do well on different parts of a problem. Tournament selection, on the other hand, computes a single fitness value aggregating a program's performance across all test cases. No matter how this aggregation is performed (e.g. summed errors, implicit fitness sharing [20], etc.), it emphasizes the selection of *generalist* individuals that perform well across all test cases. An individual can achieve terrible fitness and low probability of being selected by tournament selection if it performs very poorly on a single test case, even if it has excellent performance on all other cases; such an individual would often be selected by lexicase selection.

Other variants of lexicase selection have made alterations to other parts of the algorithm. In the initial work describing lexicase selection, what is now considered standard lexicase selection was described as "global pool, uniform random sequence, elitist lexicase parent selection" [25]. Each of these areas suggests part of the algorithm that could be changed. For example, "elitist" refers to the fact that only those individuals with exactly the best error on a test case will continue. This constraint has been relaxed in epsilon lexicase selection, in which any individual with an error value within an "epsilon" of the best error value on a case will continue to the next step [17, 18]. This variant has proved very successful on continuous-error problems, for which lexicase selection had previously performed poorly.

Since the test cases at the start of the shuffled list of cases have the most impact on selection, every selection will treat some cases as more important than others, but those cases will be different in different selection events. As indicated by "uniform random sequence" above, most work has used a uniform shuffling of test cases, giving each case equal probability of appearing at any point in the shuffled order. Since the invention of lexicase selection, many researchers (the authors included) have speculated that there must be some better way to arrange the test cases than using completely uniform shuffling. In fact, Spector tested many ad hoc methods of weighting the test case shuffle around the time lexicase selection was invented, but none of them proved superior in initial testing (L. Spector, personal communication, 2012). Burks and Punch describe a variant of lexicase selection that does not use uniform shuffling of test cases, which we discuss in more detail below and use as a comparison [1].

6.3   Weighted Shuffle

In our experiments, we consider three different methods for shuffling the test cases in a non-uniform manner for lexicase selection. Each of these shuffling methods requires a technique for weighting or ordering the test cases, which we will call the *bias metric*. The bias metric, when applied to a test case, will produce the weight for that test case.

6.3.1   Shuffling Methods

Weighted shuffle first scores each test case by the chosen bias metric, assigning the result as the weight for the case. Then, a list of test cases is built by selecting cases one at a time, with higher weighted test cases having a greater chance of being selected at each step. The weighted selection can be modeled with a roulette wheel. If a test case has a higher weight on the bias metric, its slice of the roulette wheel is larger. We then randomly select a test case based on these slices. This process is repeated until we have a weighted ordering of the test cases. This Weighted shuffle is repeated for every parent selected during a generation, meaning that different orderings will occur during the generation, but they will all use the same weights when performing the shuffle.

This Weighted shuffle algorithm, as far as we can tell, is a standard method for performing weighted shuffle. For example, this is the weighted shuffle implemented in Haskell [2].

Ranked shuffle takes the test cases and ranks them by the selected bias metric. Ranked shuffle then selects a random integer upper bound, uniformly selected between 1 and the number of test cases inclusive. Next, another uniform random integer is selected between 1 and that upper bound, inclusive; this is the index of the chosen test case. The test case at this index becomes the first test case in the new shuffled order. This same process then repeats for the remaining cases, adding each selected case to the end of the list so far. With this method, the test cases with a better rank (ex. 1, 2, 3, ...) are more likely to be chosen at each step because they are more likely to be within the range from 1 to the selected upper bound. The motivation for Ranked shuffle is that the chance of being selected is based on rank, instead of weight, and thus will not be as skewed by large differences in weight.

During each step of the Ranked Shuffle process, we choose a case out of T test cases. The case with rank $t \in 1, \ldots, T$ has probability of being selected of

$$P(t) = \frac{1}{T} \sum_{i=t}^{T} \frac{1}{i},$$

which can be seen because it will have $1/i$ chance of being chosen for each index $i \geq t$. This distribution is "a discretized version of the negative log distribution" [3], and for every integer $t \in 1, \ldots, T$, is equivalent to

$$P(t) = \frac{-\log(t/T)}{T}.$$

Fixed-order lexicase-based tournament selection (FOLBaT) is what we will call a variant of lexicase selection introduced by Burks and Punch that does not use uniform shuffling of test cases [1]. In fact, they use a fixed ordering of the test cases for every selection in a generation, instead of shuffling the test cases at all. They base this ordering on how well the population performs on the test cases that generation, with more difficult test cases coming first. Since the test case ordering is fixed each generation, if the entire population were used in each selection, the same exact individual would be selected every time. Instead, this method only applies lexicase to a subset of the population, as in tournament selection. Thus we will call this method fixed-order lexicase-based tournament selection.

For each generation, the test cases are ordered deterministically for every selection (although ties are broken randomly for each selection). The original work using FOLBaT selection uses test case orderings sorted by two different bias metrics: the Number-of-Nonzeros and Average metrics described below [1]. We use tournament size of 7, and experiment with using other bias metrics as well.

6.3.2   Bias Metrics

Some of our bias metrics tend to order "easier" test cases earlier, some order "harder" test cases earlier, and some base the ordering on the variance of the population error values on the cases. The Number-of-Zeros metric counts the number of individuals in the population that achieve zero (i.e. perfect) error on the given test case. This means that easier test cases that the population tends to get correct more often are given more weight, and therefore tend to appear earlier when shuffled. The Number-of-Zeros-Inverse metric simply divides 1 by the Number-of-Zeros metric. Thus, the weights are inverted, and more difficult test cases will be more likely to appear earlier when shuffled.

Similarly, the Number-of-Nonzeros metric counts the number of individuals in the population that do not achieve zero error on the test case. Thus it orders harder cases earlier. Note that this weighting is not equivalent to the Number-of-Zeros-Inverse weighting, since the relative weights will be different between test cases. As we will see below, this difference is not simply theoretical, since these methods give significantly different results in our empirical tests. We also try a Number-of-Nonzeros-Inverse metric that, as above, divides 1 by the Number-of-Nonzeros metric.

We could also imagine that there might be more information in the actual error values for each test case, not just whether an individual perfectly passes the case or not. Thus, we use a Median metric, which uses the median error in the population on a test case as its weight. In this setting, a higher median error will give more weight to the test case, so harder cases will come earlier. We also test a Median-Inverse metric, with which easier cases will come first. We also use an Average error metric, though we do worry that outliers may make some test cases dominate the weighting. The original FOLBaT paper used Average, so here it serves as a comparison metric [1]. Again, higher average error will give more weight, so harder cases will come earlier.

Finally, we also could imagine that it would be useful to have cases that differentiate more between individuals to come earlier; thus, we also try a Variance metric, which uses the variance of errors on a test case as its weight. Thus cases that have more varied errors will come earlier in the ordering. To be thorough, we also include a Variance-Inverse metric, where cases that have less divergent errors tend to come first.

6.4 Experimental Setup

We conducted experiments to compare our weighted shuffle lexicase selection variants to regular lexicase selection. Below we describe the experiments, including the problems and GP system we used.

6.4.1 Problems

For our experiments, we use two general program synthesis problems from a recent benchmark suite [8]. The problems in this suite, which are taken from introductory programming textbooks, require a range of data types and programming constructs to solve. We chose problems for which lexicase selection has performed well but has also showed room for improvement, so that we can expect important differences in performance to be visible. The first problem, Replace Space With Newline (RSWN), requires a program to take a string as input and print the string after replacing all spaces in the input with newline characters. It also requires the program to functionally return an integer representing the number of non-whitespace characters in the input. The second problem, Syllables, also gives a string as input. The program must count the number of vowels in the string, and then print that number as X in the string "The number of syllables is X".

In each of our experiments, we report the number of successful programs out of 100 runs. Here, a program must pass both the test cases used during evolution as well as an unseen test set in order to be called a solution. We created the both data sets using the methods described with the benchmark suite [8]. We will also plot the median *behavioral diversity* of populations across sets of runs, which is defined as the proportion of distinct behavior vectors of individuals in the population [11]. Here, a behavior vector is the list of outputs of a program when run on the test cases.

Table 6.1 PushGP
parameters used in our
experiments

Parameter	Value
Runs per problem/parameter combination	100
Population size	1000
Maximum generations	300
Genetic operator	*Prob*
Alternation	0.2
Uniform mutation	0.2
Uniform close mutation	0.1
Alternation followed by uniform mutation	0.5

6.4.2 Push and PushGP

For our experiments, we use the PushGP system, which has previously been used
extensively on the benchmark problems we use here [4, 5, 7, 8, 21]. PushGP
evolves programs in the Push programming language, a stack-based language
designed specifically for GP [24, 26]. Push has many features that make it well-
suited for general-purpose program synthesis, such as the availability of many data
types and control-flow constructs. Besides the language it evolves programs in,
PushGP is otherwise a standard generational GP system. For this work, we use the
Clojure implementation of PushGP, which is currently the most actively-developed
implementation.[1]

We give the PushGP parameters that we use in our experiments in Table 6.1.
Our experiments use Plush genomes, the linear genome representation of Push
programs [10]. The genetic operators in Table 6.1 act on these Plush genomes.
Alternation is a crossover of two parents, and uniform mutation and uniform close
mutation act on one parent; more details can be found in [10].

6.5 Results

We present the number of successful runs out of 100 for each setting on the Replace
Space With Newline (RSWN) problem in Table 6.2 and the Syllables problem
in Table 6.3. As a comparison, regular lexicase found 54 successful programs
on RSWN and 22 successful programs on Syllables. The success results in these
tables show that none of the combinations of shuffle methods with bias metrics
significantly improve performance compared to regular lexicase selection. In fact,
some give significantly worse results, using a pairwise chi-square test with Holm
correction for multiple comparisons.

---

[1] https://github.com/lspector/Clojush.

Table 6.2 Number of successes out of 100 runs on the Replace Space With Newline problem

Type	Bias metric	Weighted	Ranked	FOLBaT
Easy-first	Number-of-zeros	<u>13</u>	40	<u>6</u>
	Number-of-nonzeros-inverse	54	43	<u>7</u>
	Median-inverse	53	39	<u>4</u>
Hard-first	Number-of-zeros-inverse	52	49	35
	Number-of-nonzeros	61	44	40
	Median	50	53	<u>26</u>
	Average	33	45	<u>17</u>
Variance-based	Variance	30	57	<u>11</u>
	Variance-inverse	52	53	30

Underlined results are significantly worse than regular lexicase selection, which produced 54 successes. No results were significantly better than regular lexicase

Table 6.3 Number of successes out of 100 runs on the Syllables problem

Type	Bias metric	Weighted	Ranked	FOLBaT
Easy-first	Number-of-zeros	20	12	7
	Number-of-nonzeros-inverse	13	10	8
	Median-inverse	19	12	8
Hard-first	Number-of-zeros-inverse	11	16	<u>2</u>
	Number-of-nonzeros	17	14	<u>3</u>
	Median	20	17	6
	Average	14	15	5
Variance-based	Variance	11	13	20
	Variance-inverse	16	19	10

Underlined results are significantly worse than regular lexicase selection, which produced 22 successes. No results were significantly better than regular lexicase

While we tried every bias metric with each shuffle method, some combinations seem more relevant to consider than others. For example, in the paper describing FOLBaT, the authors use the Number-of-Nonzeros and Average bias metrics [1]. Our results with FOLBaT are mixed for these metrics. In fact, we expected the Weighted and Ranked methods to perform poorly with the Average bias metric, since we imagined it could be heavily skewed by outliers. The results show that while neither performed exceptionally well with Average, neither did exceptionally poorly either.

With the Ranked shuffle and FOLBaT, two sets of two methods should produce equivalent rankings of cases and therefore comparable results. That is, Number-of-Zeros and Number-of-Nonzeros-Inverse should behave identically, since counting the number of zeros will produce the same ordering of test cases as taking the inverse of the number of nonzeros. Similarly, Number-of-Zeros-Inverse and Number-of-Nonzeros should also behave equivalently with Ranked shuffle. As expected, the numbers of successes for each of these combinations is not significantly different

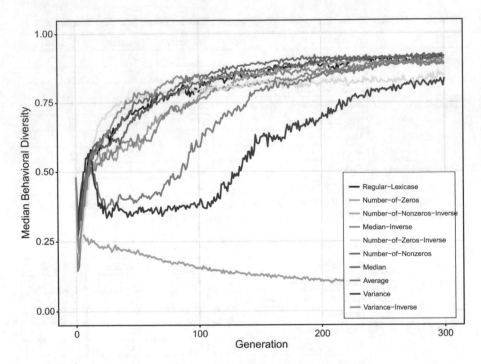

Fig. 6.1 For **Weighted shuffle**, the average population behavioral diversity of each bias metric plotted over the generations of each set of runs on the RSWN problem. Note that the black Regular-Lexicase line is mostly hidden behind the red Median line

from one another. This equivalency does not hold for Weighted shuffle, where the relative differences in weight matter for the shuffle.

Figure 6.1 plots the average population behavioral diversity for each bias metric when using Weighted shuffle on the RSWN problem. This plot shows that Weighted shuffle is not able to produce significantly higher levels of behavioral diversity than regular lexicase selection, no matter what bias metric is used. While many of the bias metrics produce similar diversity to regular lexicase, a few result in significantly worse diversity. This is especially apparent with Number-of-Zeros, Variance, and Average, the three bias metrics that performed worst on this problem, showing a correlation between poor performance and poor diversity.

Figure 6.2 gives the same diversity plots, except for the Ranked shuffling method. Here, we see diversity more akin to that of regular lexicase, though many of the bias metrics have lower diversity in the first 150 generations of runs. There does not seem to be much correlation between diversity and success rate, with some of the metrics that create lower diversity still finding comparable numbers of solutions.

Finally, we give the diversity results for FOLBaT in Fig. 6.3. Most of the bias metrics we used with FOLBaT do not exhibit the ability to increase and maintain diversity shown by regular lexicase selection. The two exceptions are with the Median and Average bias metrics; these metrics achieve high levels of diversity,

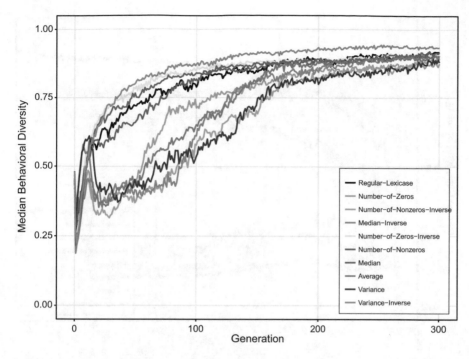

Fig. 6.2 For **Ranked shuffle**, the average population behavioral diversity of each bias metric plotted over the generations of each set of runs on the RSWN problem

although still lower than with regular lexicase. Interestingly, both of these metrics performed poorly in success rates, while some metrics with lower levels of diversity performed significantly better.

6.6   Discussion

Both Weighted shuffle and Ranked shuffle perform about as well as regular lexicase for most of the bias metrics. Weighted shuffle had slightly better results than Ranked shuffle with the bias metrics that gave the best results, but also much worse results with the worst bias metrics. FOLBaT, on the other hand, was often significantly worse than regular lexicase, with every bias metric except Variance-Inverse giving significantly worse results on one of the two problems.

As for the bias metrics, none stand out as particularly good or bad on these two problems, at least when only considering Weighted and Ranked shuffles. In fact, some of the bias metrics that give the worst results on one problem give the best results on the other problem.

The success rate results show that none of our combinations of shuffle methods with bias metrics resulted in significantly better results than with regular lexicase,

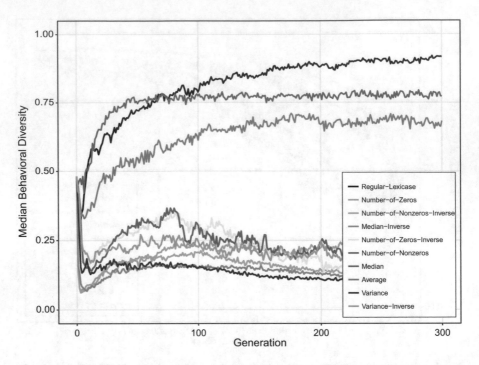

Fig. 6.3 For **FOLBaT**, the average population behavioral diversity of each bias metric plotted over the generations of each set of runs on the RSWN problem

which does not weight the shuffling of test cases. Thus it is difficult to recommend any of these shuffle methods or bias metrics over regular lexicase selection. Why, we must ask, does this intuitive idea of weighting the shuffling of test cases not lead to improvements?

Although we do not have any quantitative data, we have seen anecdotal evidence that suggests that with Weighted shuffle, some of the bias metrics that perform best often use near-uniform distributions of test cases when shuffling. Test cases have similar or identical weights when most individuals in the population perform similarly on them. When most or all test cases have similar wights, Weighted shuffle acts very similarly to the uniform shuffle used by regular lexicase selection. Thus, it is not surprising that bias metrics that use near-uniform shuffling of test cases will give good performance results similar to regular lexicase selection. What is surprising is that many of the bias metrics that produce more weighting of test cases perform poorly, suggesting simply that weighting the shuffle leads to poor results. Note that these anecdotes only apply to Weighted shuffle, since the other shuffle methods result in different distributions of shuffles.

One possible explanation for the significantly worse results that we do see with some bias metrics with Weighted shuffle, and the slightly worse results for Ranked shuffle, is that they over-concentrate on some test cases while ignoring others. Such behavior may reduce the population diversity by limiting which test cases influence selection.

Turning attention to the diversity figures, we note that maintaining high diversity is somewhat correlated to better performance, but not always. In some cases, low diversity may play a large part in poor performance, especially with FOLBaT. But, this correlation does not always hold; for example, with Ranked shuffle, the Variance bias metric had the highest number of success, yet exhibited some of the lowest diversity in the first half of runs.

Of the three shuffle methods, FOLBaT performed the worst and also the exhibited the lowest diversity. We believe this lack in diversity is caused by FOLBaT not using different orderings of test cases. This feature of lexicase selection allows it to emphasize different test cases with each selection, and therefore select many different individuals that perform well on different tests. Since FOLBaT doesn't shuffle the test cases at all, it always emphasizes the same cases within a generation. FOLBaT's reliance on a single sorting of the test cases gives a total ordering of the population, which can be seen as a scalar fitness value assigned to each individual based on rank. It only selects different individuals because it uses a tournament, and therefore the best individuals for each generation's ordering will not be present in every tournament. Avoiding a scalar fitness value is one of the key tenants of lexicase selection. This use of a scalar fitness value, a characteristic that it shares with tournament selection, may explain FOLBaT's inability to increase and maintain diversity, similar to tournament selection [6].

What does this teach us about lexicase selection? Lexicase selection's ability to emphasize different test cases with each selection seems paramount to its ability to maintain diversity, by selecting a wide range of individuals that specialize in different combinations of test cases. Using non-uniform shuffle decreases this uniformly random aspect of lexicase selection, which seems to have a neutral or negative impact on results.

Additionally, lexicase selection already places emphasis on individuals that uniquely perform well on single or multiple test cases, especially if such individuals also perform well on other cases. It appears to not be useful to place extra emphasis on some test cases; such favoring does not add utility to how often those cases appear early in the shuffle, and other times could lead to other cases being under-emphasized.

6.7   Related Work

The primary idea behind this work, that some test cases should be emphasized more than others based on how well the population performs on them, shares motivation with other parent selection methods. Each of the following methods uses tournament selection, but modifies fitness calculations in some way. In implicit fitness sharing (IFS), fitness is weighted so that test cases that are solved by fewer individuals receive more weight [20]. On problems such as those in this paper where test cases are non-binary, it is necessary to use a non-binary adaptation of IFS [14]. Earlier work showed that this non-binary IFS produced significantly worse results than lexicase selection on the two problems presented here, and on the benchmark suite that the problems come from more generally [8]; it has also been shown to produce lower levels of population diversity [5].

In "co-solvability" fitness, IFS has been extended to not weight individual cases, but instead pairs of test cases [13]. In this way, individuals that solve pairs of test cases not often solved by other individuals receive more reward. "Historically-assessed hardness" is another generalization of IFS to non-binary test cases, where fitness on each test case is scaled based on the performance of the population [12].

6.8 Conclusions and Future Work

This work demonstrates that using weighted shuffle of the test cases in lexicase selection does not increase the performance of GP. Over a wide array of methods for biasing the shuffle, including some that emphasize easy test cases and some that emphasize difficult test cases, we do not see any significant gains in performance compared to regular lexicase selection. We note a correlation between success rate and the ability to maintain diversity across some of our experimental results, in which methods that produced lower diversity were also those with the worst performance, though this correlation does not hold across the board.

These results, while discouraging with regard to improving performance, do suggest that it is not necessary to use any additional test case shuffling scheme in order to achieve good results with lexicase selection. Thus we recommend the continued use of lexicase selection with uniform random shuffling.

We hypothesize that one potential problem with the shuffle methods we present here is that they over-emphasize certain cases, which over-selects specific members of the population. In the future, we could consider whether there are ways to not over-emphasize specific test cases while still performing weighted shuffling. Such a scheme may be able to achieve better performance results while maintaining high levels of diversity. On the other hand, the resulting shuffles will be more similar to uniform shuffling, and therefore may simply behave more similarly to regular lexicase selection.

Acknowledgements Hammad Ahmad, Lee Spector, and Nicholas Freitag McPhee shared interesting discussions that were very helpful in conducting this work.

References

1. Burks, A.R., Punch, W.F.: An investigation of hybrid structural and behavioral diversity methods in genetic programming. In: Genetic Programming Theory and Practice XIV, Genetic and Evolutionary Computation. Springer, Ann Arbor, USA (2016), in press (2018)
2. Data.random.shuffle.weighted. https://hackage.haskell.org/package/random-extras-0.19/docs/Data-Random-Shuffle-Weighted.html. Accessed: 2017-05-01

3. Drury, M.: Does this discrete distribution have a name? Cross Validated. URL https://stats.stackexchange.com/q/152786. Accessed: 2017-11-25

4. Helmuth, T.: General program synthesis from examples using genetic programming with parent selection based on random lexicographic orderings of test cases. Ph.D. dissertation, University of Massachusetts, Amherst (2015). URL http://scholarworks.umass.edu/dissertations_2/465/

5. Helmuth, T., McPhee, N.F., Spector, L.: Lexicase selection for program synthesis: A diversity analysis. In: Genetic Programming Theory and Practice XIII, Genetic and Evolutionary Computation, pp. 151–167. Springer, Ann Arbor, USA (2015). https://doi.org/10.1007/978-3-319-34223-8. URL http://cs.wlu.edu/~helmuth/Pubs/2015-GPTP-lexicase-diversity-analysis.pdf

6. Helmuth, T., McPhee, N.F., Spector, L.: Effects of lexicase and tournament selection on diversity recovery and maintenance. In: GECCO '16 Companion: Proceedings of the Companion Publication of the 2016 Annual Conference on Genetic and Evolutionary Computation, pp. 983–990. ACM, Denver, Colorado, USA (2016). https://doi.org/10.1145/2908961.2931657

7. Helmuth, T., McPhee, N.F., Spector, L.: The impact of hyperselection on lexicase selection. In: T. Friedrich (ed.) GECCO '16: Proceedings of the 2016 on Genetic and Evolutionary Computation Conference, pp. 717–724. ACM, Denver, USA (2016). https://doi.org/10.1145/2908812.2908851

8. Helmuth, T., Spector, L.: General program synthesis benchmark suite. In: GECCO '15: Proceedings of the 2015 on Genetic and Evolutionary Computation Conference, pp. 1039–1046. ACM, Madrid, Spain (2015). http://doi.acm.org/10.1145/2739480.2754769

9. Helmuth, T., Spector, L., Matheson, J.: Solving uncompromising problems with lexicase selection. IEEE Transactions on Evolutionary Computation 19(5), 630–643 (2015). https://doi.org/10.1109/TEVC.2014.2362729

10. Helmuth, T., Spector, L., McPhee, N.F., Shanabrook, S.: Linear genomes for structured programs. In: Genetic Programming Theory and Practice XIV, Genetic and Evolutionary Computation. Springer, Ann Arbor, USA (2016), in press (2018)

11. Jackson, D.: Promoting phenotypic diversity in genetic programming. In: PPSN 2010 11th International Conference on Parallel Problem Solving From Nature, *Lecture Notes in Computer Science*, vol. 6239, pp. 472–481. Springer, Krakow, Poland (2010). https://doi.org/10.1007/978-3-642-15871-1_48

12. Klein, J., Spector, L.: Genetic programming with historically assessed hardness. In: Genetic Programming Theory and Practice VI, Genetic and Evolutionary Computation, chap. 5, pp. 61–75. Springer, Ann Arbor (2008). https://doi.org/10.1007/978-0-387-87623-8_5

13. Krawiec, K., Lichocki, P.: Using co-solvability to model and exploit synergetic effects in evolution. In: PPSN 2010 11th International Conference on Parallel Problem Solving From Nature, *Lecture Notes in Computer Science*, vol. 6239, pp. 492–501. Springer, Krakow, Poland (2010). https://doi.org/10.1007/978-3-642-15871-1_50

14. Krawiec, K., Nawrocki, M.: Implicit fitness sharing for evolutionary synthesis of license plate detectors. In: Applications of Evolutionary Computing, EvoApplications 2012, *Lecture Notes in Computer Science*, vol. 7835, pp. 376–386. Springer, Vienna, Austria (2013). https://doi.org/10.1007/978-3-642-37192-9_38

15. Krawiec, K., O'Reilly, U.M.: Behavioral programming: A broader and more detailed take on semantic GP. In: Proceedings of the 2014 Annual Conference on Genetic and Evolutionary Computation, GECCO '14, pp. 935–942. ACM, New York, NY, USA (2014). http://doi.acm.org/10.1145/2576768.2598288

16. Krawiec, K., Swan, J., O'Reilly, U.M.: Behavioral program synthesis: Insights and prospects. In: Genetic Programming Theory and Practice XIII, Genetic and Evolutionary Computation Series, pp. 169–183. Springer (2015)

17. La Cava, W., Moore, J.: A general feature engineering wrapper for machine learning using epsilon-lexicase survival. In: M. Castelli, J. McDermott, L. Sekanina (eds.) EuroGP 2017: Proceedings of the 20th European Conference on Genetic Programming, *LNCS*, vol. 10196, pp. 80–95. Springer Verlag, Amsterdam (2017). https://doi.org/10.1007/978-3-319-55696-3_6

18. La Cava, W., Spector, L., Danai, K.: Epsilon-lexicase selection for regression. In: T. Friedrich (ed.) GECCO '16: Proceedings of the 2016 Annual Conference on Genetic and Evolutionary Computation, pp. 741–748. ACM, Denver, USA (2016). https://doi.org/10.1145/2908812.2908898
19. Liskowski, P., Krawiec, K., Helmuth, T., Spector, L.: Comparison of semantic-aware selection methods in genetic programming. In: C. Johnson, K. Krawiec, A. Moraglio, M. O'Neill (eds.) GECCO 2015 Semantic Methods in Genetic Programming (SMGP'15) Workshop, pp. 1301–1307. ACM, Madrid, Spain (2015). http://doi.acm.org/10.1145/2739482.2768505
20. McKay, R.I.: Fitness sharing in genetic programming. In: Proceedings of the Genetic and Evolutionary Computation Conference (GECCO-2000), pp. 435–442. Morgan Kaufmann, Las Vegas, Nevada, USA (2000)
21. McPhee, N.F., Finzel, M., Casale, M.M., Helmuth, T., Spector, L.: A detailed analysis of a PushGP run. In: Genetic Programming Theory and Practice XIV, Genetic and Evolutionary Computation. Springer, Ann Arbor, USA (2016), in press (2018)
22. McPhee, N.F., Ohs, B., Hutchison, T.: Semantic building blocks in genetic programming. In: Proceedings of the 11th European Conference on Genetic Programming, EuroGP 2008, *Lecture Notes in Computer Science*, vol. 4971, pp. 134–145. Springer, Naples (2008)
23. Moraglio, A., Krawiec, K., Johnson, C.G.: Geometric semantic genetic programming. In: Parallel Problem Solving from Nature, PPSN XII (part 1), *Lecture Notes in Computer Science*, vol. 7491, pp. 21–31. Springer, Taormina, Italy (2012)
24. Spector, L.: Autoconstructive evolution: Push, PushGP, and Pushpop. In: Proceedings of the Genetic and Evolutionary Computation Conference (GECCO-2001), pp. 137–146. Morgan Kaufmann, San Francisco, California, USA (2001). URL http://hampshire.edu/lspector/pubs/ace.pdf
25. Spector, L.: Assessment of problem modality by differential performance of lexicase selection in genetic programming: a preliminary report. In: Proceedings of the fourteenth international conference on Genetic and evolutionary computation conference companion, GECCO Companion '12, pp. 401–408. ACM, New York, NY, USA (2012). https://doi.org/10.1145/2330784.2330846
26. Spector, L., Klein, J., Keijzer, M.: The Push3 execution stack and the evolution of control. In: GECCO 2005: Proceedings of the 2005 conference on Genetic and evolutionary computation, pp. 1689–1696. ACM Press, Washington DC, USA (2005). https://doi.org/10.1145/1068009.1068292. URL http://www.cs.bham.ac.uk/~wbl/biblio/gecco2005/docs/p1689.pdf

Chapter 7
Relaxations of Lexicase Parent Selection

Lee Spector, William La Cava, Saul Shanabrook, Thomas Helmuth, and Edward Pantridge

Abstract In a genetic programming system, the parent selection algorithm determines which programs in the evolving population will be used as the material out of which new programs will be constructed. The lexicase parent selection algorithm chooses a parent by considering all test cases, individually, one at a time, in a random order, to reduce the pool of possible parent programs. Lexicase selection is ordinarily strict, in that a program can only be selected if it has the best error in the entire population on the first test case considered, and the best error relative to all other programs that remain in the pool each time it is reduced. This strictness may exclude high-quality candidates from consideration for parenthood, and hence from exploration by the evolutionary process. In this chapter we describe and present results of four variants of lexicase selection that relax these strict constraints: epsilon lexicase selection, random threshold lexicase selection, MADCAP epsilon lexicase selection, and truncated lexicase selection. We present the results of experiments with genetic programming systems using these and other parent selection algorithms on symbolic regression and software synthesis problems. We also briefly discuss the relations between lexicase selection and work on many-objective optimization,

L. Spector (✉)
Hampshire College, Amherst, MA, USA
e-mail: lspector@hampshire.edu

W. La Cava
Institute for Biomedical Informatics, University of Pennsylvania, Philadelphia, PA, USA
e-mail: lacava@upenn.edu

S. Shanabrook
University of Massachusetts, Amherst, MA, USA

T. Helmuth
Hamilton College, Clinton, NY, USA
e-mail: thelmuth@hamilton.edu

E. Pantridge
MassMutual, Amherst, MA, USA
e-mail: EPantridge@MassMutual.com

© Springer International Publishing AG, part of Springer Nature 2018
W. Banzhaf et al. (eds.), *Genetic Programming Theory and Practice XV*,
Genetic and Evolutionary Computation, https://doi.org/10.1007/978-3-319-90512-9_7

and the implications of these considerations for future work on parent selection in genetic programming.

7.1 Introduction

Almost all parent selection algorithms used in genetic programming systems involve not only comparisons among potential parent programs, but also random choices.

For example, in fitness-proportionate selection, we simulate the spinning of a roulette wheel, with the size of the pocket for each parent being inversely proportional to its total error over the set of test cases. It is possible for any program in the population to be selected as a parent, and the choice among potential parents, while random, is biased so that higher quality programs have higher probabilities of being selected.

In tournament selection, we first choose a tournament set randomly from the population, with each program in the population having equal probability of being chosen. We then select the best program in the tournament set to serve as the parent, where the "best" program is the one with the lowest total error. Here random choices are made first, to determine the tournament set, followed by a choice that is driven by the quality of the programs that have been chosen to participate in the tournament.

One could, in principle, avoid making any random choices in parent selection, but this is rarely done. For example, one could use a form of pure elitism, in which each program in the "best" $n\%$ of the population is selected as a parent some pre-specified number of times.

In lexicase selection, randomization plays a particularly central role, but rather than being applied directly to choices of programs, it is applied to sequences of individual test cases by which programs are compared. In lexicase selection, parents are selected through a filtering process that is iterated over a sequence of test cases that is randomly shuffled for each parent selection event. Once randomly re-ordered, the test cases are considered one at a time, with only the best programs for each case retained at the corresponding filtering step.

Tournament selection can be thought of as subjecting a subset of the potential parents to a challenge of this form for each selection event: "Are you better, overall, than these other randomly-chosen programs?" By contrast, lexicase selection can be thought of as subjecting each potential parent to a sequence of challenges—"Are you the best on *this* test case? And of those of you who are, are you best on *this* one? Etc."—among which the order is randomized.

Lexicase selection has been shown to be advantageous in several contexts. It often allows problems to be solved more quickly and reliably than they can be without it [7, 17], and in some cases allows for the solution of problems that cannot otherwise be solved at all [6].

The power of lexicase selection appears to stem from the way in which it leverages multiple, randomized challenges to guide search. The randomization of test case order allows the parent selection process to be sensitive to more information about the strengths and weaknesses of programs in the population than it can be under the approach used in tournament selection. In fact, recent experiments with weighted shuffling of test cases produced similar or worse results, suggesting that the uniform shuffling of test cases allows lexicase selection to better sample useful programs in the population [22]. This randomization of challenges allows lexicase selection to be sensitive not only to the performance of programs on all test cases considered in aggregate, but also to their performance on all subsets of the test cases; in this way, lexicase selection often selects individuals that specialize in some test cases while performing poorly on others. Considering all subsets of test cases explicitly would require exponential resources, but randomization allows them to be considered implicitly, through random sampling.

When a parent selection algorithm is sensitive to more information about the strengths and weaknesses of programs, then that information may be used to provide better guidance to evolutionary search in different ecological circumstances. Semantic- or behavior-aware genetic programming methods (such as lexicase selection) take into account information about a program's execution or its individual outputs/errors on test cases, going beyond methods that simply use a single fitness value [8, 9, 15, 18].

Might variations of lexicase selection perform even better on problems of specific kinds? In this chapter we describe and present data on four "relaxed" forms of lexicase selection, each of which allows some programs to be selected that would not be selected by ordinary lexicase selection; these are epsilon lexicase selection, random threshold lexicase selection, MADCAP epsilon lexicase selection, and truncated lexicase selection. Among the motivations for considering these forms of relaxation is the hypothesis that ordinary lexicase selection can sometimes be too strict, insofar as it eliminates any opportunity to serve as a parent for some programs that are quite good in many respects.

In the following sections, we first describe the most basic form of lexicase selection, on which the other selection methods described in this chapter are based. We then describe each of the four relaxed versions of lexicase selection in turn. Following the descriptions of the algorithms, we present and discuss the results of experiments involving all of the described algorithms, along with a few others from the literature to facilitate broader comparisons. These experiments involve eight symbolic regression problems and five software synthesis problems. We conclude with a brief discussion of the relation between work on lexicase selection and work on many-objective optimization, and we discuss the implications of our results for future research.

7.2 Lexicase Selection

Lexicase selection is designed for problems in which candidate solutions are assessed with multiple test cases. In most other parent selection methods, a candidate solution's performance over multiple test cases is aggregated into a single measure, for example, an average error value, and this single aggregate measure is used as the basis of selection. In lexicase selection, no aggregation is performed; the measures for each individual test case are retained, and they may all be used, individually, in the parent selection process.

Although lexicase selection can be used for other kinds of performance measures as well, for the sake of simplicity we will refer to measures of performance on individual test cases as "errors," and we will assume that we are seeking a solution that minimizes all errors.

With the most basic form of lexicase selection ("global pool, uniform random sequence, elitist lexicase parent selection" [20], which will also refer to below as "ordinary lexicase selection"), when the genetic programming system requires a parent to use for the production of offspring, we first shuffle a sequence of the test cases that are being used to assess programs in the population. We then form a pool that initially contains all of the programs in population. We will winnow this pool down to a single selected program by considering each test case in turn. When each test case is considered, we first note the lowest error that any program in the pool has for that test case. We then eject all programs that have a higher error on that test case from the pool. If these ejections ever reduce the pool to a single program, then we return that program as the selected parent. If instead, we exhaust the test cases and still have more than one program in the pool, then we return one of them randomly.

Why might one expect this selection method to be useful? One reason is that it allows programs to be selected if they perform particularly well on individual test cases, or on collections of test cases, even if they perform poorly on many others. This allows specialists to produce offspring that may build on their specialties, perhaps in conjunction with other specialties that they may have inherited from other ancestors. The full reasons for lexicase selection's utility, however, are more complex, and still under investigation [3–5, 15].

A variety of time optimizations of lexicase selection are possible. For example, we can include in the initial pool just a single random representative of any group that shares the same errors for all test cases. Doing so will decrease the number of programs in the pool that will have to be filtered, and it will also allow fewer test cases to be considered for some parent selection events.

7.3   Epsilon Lexicase Selection

In prior work, it was noted that lexicase selection would sometimes perform poorly on symbolic regression problems involving floating-point errors [7]. It was thought that this was due to the fact that in these contexts it would often be the case that most or all programs in the population would have unique error values when any particular test case is considered. In such situations, the strictness of lexicase selection would reduce the candidate pool to a single program as soon as the first test case is considered, and parenthood decisions would often be made on the basis of single test cases. These considerations led to the development of epsilon lexicase selection, which has indeed proven to be useful for problems involving floating-point errors [11, 12].

In epsilon lexicase selection we relax the elitism of the filtering steps. Rather than retaining only the programs with exactly the lowest error on the current test case, we retain all programs that are "close enough"—that is, those with errors within some small *epsilon* of the lowest error of any program in the pool on the current test case.

While the reasons for epsilon lexicase selection's good performance are still under investigation, an intuitive case for its success can be based on the considerations sketched above. In a population in which no two programs have the same error for any test case, which is not terribly hard to imagine for problems with floating-point errors, ordinary lexicase selection would select every parent based on a single test case. Specialists would still be selected, but not programs that perform well on multiple test cases. By allowing programs with errors that are "close enough" to the minimum error on a test case to pass through the filter, the algorithm will once again be able to select programs based on performance on larger subsets of the test cases. Support for this theory has been demonstrated by observing that epsilon lexicase selection uses more cases per selection event than ordinary lexicase selection does on regression problems [11, 12].

How should the epsilon in epsilon lexicase selection be determined? Several approaches to this question have been explored, with the most consistently good performance having been obtained so far with a method dubbed "MAD" epsilon lexicase selection, for "Median Absolute Deviation from the median." Here epsilon is computed one per generation, for each test case, on the basis of all the errors for the test case across the population. Specifically, epsilon for a particular test case is computed as the median of the differences between errors on the case and the median error for the case. When we use the name "epsilon lexicase selection" below, without further qualification, we are referring to this method.

7.4 Random Threshold Lexicase Selection

Epsilon lexicase selection sets a threshold for each challenge that a program must meet in order to survive a filtering step: if the program has an error for the case that differs from the best error by the threshold or less, then it survives. The threshold is set on the basis of the distribution of errors for the test cases in the population; for example, it is the median absolute deviation from the median error when the MAD version of epsilon lexicase selection is used.

The idea behind random threshold lexicase selection is to randomize the setting of the threshold as well. One motivation for doing this is the observation that the threshold in epsilon lexicase selection can be quite sensitive to changes in the distribution of error values across the population. If the distribution is not sufficiently well behaved, for example because of unusual features of the problem that we are trying to solve, or because high error penalties are imposed on programs that violate specified constraints, or because the genetic operators being used often produce large changes in error between parent and child, then one might expect the thresholds used by epsilon selection to be unhelpful.

For this reason, random threshold lexicase selection was developed to choose thresholds that are derived from the errors present in the population, but less sensitive to their specific distributions. Specifically, at each step of filtering, we choose an error randomly from those present in the current pool for the current case. We then retain only those programs that have the chosen error or better for the current case. If the randomly selected error happens to be the best error in the pool, then the filtering at this step will be equivalent to that used by ordinary lexicase selection. If it happens to be the worst error in the pool, then no filtering at all will take place for the current test case in the current selection event.

One can think of random threshold lexicase selection as randomly sampling combinations of relative tightness of selection on different test cases, all within lexicase selection's random ordering of test cases. So there is a sense in which all orderings of test cases and also all combinations of strictness vs. laxness for each test case may be considered. As with ordinary lexicase selection, however, we do not consider all of these combinations of challenges explicitly. Rather, we sample both the orderings and the strictnesses of the challenges that programs must confront.

At one extreme, when an elite error is picked at each step, this will act like ordinary lexicase selection. However, this will be rare. Consequentially, random threshold lexicase selection is a significantly relaxed form of lexicase selection, insofar as it will generally make it easier for programs to meet the challenges to survive filtering. For problems in which all errors are binary (pass/fail), it will act like lexicase selection on a random subset of the cases.

One would expect that this technique would often end up producing effects similar to those of ordinary lexicase selection, but with some test cases more-or-less skipped during some selection events, while the cases with "tight" bounds on errors will be the ones that do the major culling. How much of an effect this will have can be expected to depend on how many intermediate values there are between the elite values and the worst values; if there are many, then we might expect its effects to be quite different from those of ordinary lexicase selection.

We can think of both epsilon lexicase selection and random threshold lexicase selection as loosening lexicase selection's elitist filtering condition, and thereby weakening the challenge presented by each test case. Such weakening will generally lessen the selection pressure exerted by individual test cases while increasing the selection pressure exerted by groups of test cases that are adjacent in random shuffles.

We would generally expect random threshold lexicase selection to weaken test case challenges more than MAD epsilon lexicase selection does, since we would expect the bound provided by epsilon to be relatively tight, so that randomly chosen errors would not usually fall within it. It is possible that this will mean that random threshold lexicase selection will not provide enough selection pressure for good performance on individual test cases. Whether or not this will actually be the case is an empirical question, best answered by experiments.

7.5 MADCAP Epsilon Lexicase Selection

Random threshold lexicase selection relaxes test case challenges in a randomized way, but it may also be useful to consider methods that do something similar while nonetheless obeying the limit of relaxation used in epsilon lexicase selection. That is, it might be useful in some contexts to consider methods that vary in stringency between MAD epsilon lexicase selection and ordinary lexicase selection (which is strictly elitist), again using randomization to sample different strengths applied to different test cases. MADCAP epsilon lexicase selection is such a method.

At each filtering step of MADCAP epsilon lexicase selection, we sometimes retain just the best individuals on the case, and sometimes retain any individuals within epsilon of best, choosing randomly between these options for each test case. The application of epsilon lexicase's "cap" (threshold) is probabilistic.[1] Specifically, we provide a parameter for the probability of applying the MAD cap versus just retaining individuals with the best error. In the experiments described below, this parameter is set to 0.5. At each filtering step, we use this probability to determine whether to retain only those programs with exactly the best error in the pool on the current case, or whether to retain all programs with errors within the MAD epsilon of the best error. Thus the selectivity of MADCAP epsilon lexicase selection will be between that of ordinary lexicase selection and MAD epsilon lexicase selection.

The motivation for this formulation is that for some problems, it is required that solutions have errors that are actually zero, or at least equal to the lowest possible error, rather than just being low. Especially for these problems, but possibly for others as well, we would like selection to sometimes (probabilistically) distinguish between programs that have the minimum (possibly zero) error on a test case and programs that merely have low errors.

Intuitively, one might expect MADCAP epsilon lexicase selection to allow the genetic programming search process to hone in on minimal-error solutions. Whether this happens in practice will probably depend on several factors including the distribution of errors in the population, which will depend in turn on factors such as the genetic operators and rates that are being used. Epsilon lexicase selection is always sensitive to the distribution of errors across the population, while MADCAP epsilon lexicase selection will always provide some selection in favor of elites, regardless of the error distribution.

As with the other methods considered here, MADCAP epsilon lexicase selection uses sampling to consider, in the limit but not explicitly, all combinations of favoring vs. not favoring elites for each case and each combination of cases.

Again, whether or not this will actually be useful in practice is an empirical question, best answered by experiments.

---

[1] So "MADCAP" = Median Absolute Deviation from the median, Cap Applied Probabilistically.

7.6   Truncated Lexicase Selection

Truncated lexicase selection is a form of lexicase selection in which we limit the
number of cases that are considered in each parent selection event. The number (or
percentage) of the total cases that will be considered is a parameter of the method.
For example, suppose that we use truncated lexicase selection on a problem with 100
test cases and that we specify that 25% of cases will be used. Then for each parent
selection event, we will proceed initially as we do in ordinary lexicase selection, but
if we have filtered the pool using 25 test cases and it still contains multiple programs,
then we will immediately choose a random remaining member of the pool and return
it as the selected parent.

Epsilon lexicase selection, random threshold lexicase selection, and MADCAP
epsilon lexicase selection are all relaxed forms of lexicase selection in which the
constraints on selection are reduced with respect to the allowed error values for
individual cases. In ordinary lexicase selection, a program can only be selected
to serve as a parent if it is globally elite on at least one case and elite with
respect to the survivors in the selection pool as each subsequent case is considered.
In epsilon lexicase selection, random threshold lexicase selection, and MADCAP
epsilon lexicase selection, this requirement of eliteness is relaxed, to a greater or
lesser (and sometimes random) extent.

By contrast, in truncated lexicase selection we still require eliteness *on the test
cases that are considered*, but we place no constraints at all on the error values of the
cases that are not considered. Whether or not this form of relaxation has beneficial
impacts on the ability of the genetic programming system to solve problems, it also
has the potential to improve system runtimes by reducing the amount of computation
that must be dedicated to filtering the lexicase selection candidate pools.

7.7   Experimental Results

We include here the results from two sets of experiments on the relaxed variants of
lexicase selection presented above.

First, we present the results of comparisons of several selection methods, including ordinary lexicase selection, epsilon lexicase selection, random threshold lexicase
selection, and MADCAP lexicase selection, on a collection of eight symbolic
regression problems. For completeness, we also include comparisons to purely
random selection, tournament selection with a tournament size of 2, lasso selection
[21], age-fitness Pareto optimization [19], and deterministic crowding [16].

Table 7.1 describes the problems used for these experiments, each of which
comes from the UCI repository [14]. Table 7.2 describes the genetic programming
system parameters that were used, and also provides abbreviations for the parent
selection methods that were studied, which are used in the graph of results in

Table 7.1 Regression
problems used for method
comparisons

Problem	Dimension	Samples
Airfoil	5	1503
Concrete	8	1030
ENC	8	768
ENH	8	768
Housing	14	506
Tower	25	3135
UBall5D	5	6024
Yacht	6	309

Table 7.2 Genetic programming system settings for symbolic regression problems

Setting	Value
GP tool	ellyn
Population size	1000
Crossover/mutation	60/40%
Program length limits	[3, 50]
ERC range	[−1,1]
Generation limit	1000
Trials	50
Terminal set	{**x**, ERC, +, −, ∗, /, sin, cos, exp, log}
Elitism	Keep best
Fitness (non-lexicase methods)	MSE
Method	*Abbreviation*
Lasso [21]	lasso
Random selection	rand
Tournament selection (size 2)	tourn
Lexicase selection	lex
Age-fitness Pareto optimization [19]	afp
Deterministic crowding [16]	dc
Epsilon-lexicase selection	ep-lex
Random threshold lexicase selection	ep-lex-rand
MADCAP epsilon-lexicase selection	ep-lex-madcap

Fig. 7.1. The experiments were run using the ellyn,[2] a linear GP system described in [10] (in this experiment, no epigenetic markers were used).

As Fig. 7.1 makes clear, epsilon lexicase selection achieves the best results, and achieves the most consistently good results, across this set of problems. Ordinary lexicase selection sometimes performs reasonably well, occasionally beating competitors, but as has been noted elsewhere and motivated the development of epsilon lexicase selection in the first place, ordinary lexicase selection often performs relatively poorly in the context of floating-point errors [12].

---

[2]https://epistasislab.github.io/ellyn/.

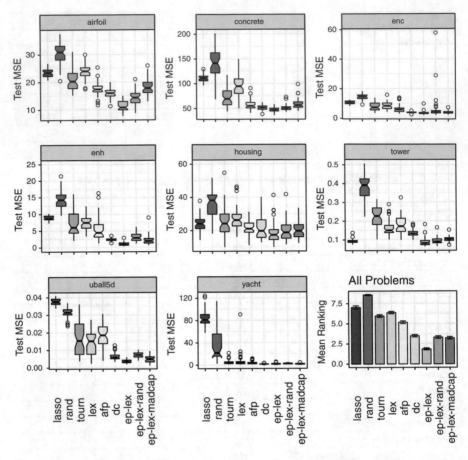

Fig. 7.1 Comparison of multiple parent selection methods on multiple symbolic regression problems. The boxplots span the upper and lower quartiles of test set mean squared error (MSE) for each problem, with a central line indicating the median. On the lower right, the mean ranking according to median MSE of each method across all problems is shown, with the error bar indicating the standard error

Random threshold lexicase selection performs much better than ordinary lexicase selection but worse than epsilon lexicase selection. Its average ranking over these regression problems is about the same as deterministic crowding. MADCAP lexicase selection has broadly similar performance, sometimes a bit better and sometimes a bit worse. It is possible, however, that each of these relaxed forms of epsilon lexicase selection will prove to be more advantageous for problems with particular characteristics that might be discovered by broadening the experiments to cover more types of problems.

In a second set of experiments, we compared ordinary lexicase selection to truncated lexicase selection on software synthesis problems from our general program synthesis benchmark suite [6]. Specifically, we conducted runs on the "Median," "Negative to Zero," "Number IO," "Replace Space with Newline,"

Table 7.3 Genetic programming system settings for software synthesis problems

Setting	Smallest	Median	Negative to zero	Number IO	RSWN	Vector average
GP tool	Clojush	Clojush	Clojush	Clojush	Clojush	Clojush
Population size	1000	1000	1000	1000	1000	1000
Generation limit	200	200	300	200	300	300
Alternation/ mutation/both	20/30/50%	20/30/50%	20/30/50%	30/20/50%	20/30/50%	20/30/50%
Max initial genome length	100	100	250	100	400	200
Test cases	100	100	200	1000	100	250

All problems are explained in detail in [6]. The full set of settings, including the set of instructions allowed in programs, can be found in the Clojush GitHub repository

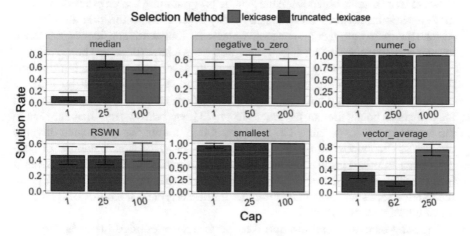

Fig. 7.2 Comparison of ordinary lexicase selection and truncated lexicase selection on software synthesis benchmark problems. The x-axis labels denote the values for the cap parameter of truncated lexicase selection in number of test cases. The values are a single test case, 25% of the total number of test cases, and 100% of the total number of test cases (which is ordinary lexicase selection). Error bars indicate the standard error

"Smallest," and "Vector Average" problems. These problems require a range of data types and programming constructs to solve. For these problems, we tested truncating lexicase selection after considering just a single test case, and after considering 25% of the test cases. We note that when only a single test cases are considered we are creating, intentionally, exactly the situation which was thought to be problematic when using ordinary lexicase selection on problems with floating-point errors, which motivated the development of epsilon lexicase selection.

The parameters for the runs comparing truncated lexicase selection with ordinary lexicase selection, using the Clojush implementation of the PushGP genetic programming system (https://github.com/lspector/Clojush), are shown in Table 7.3. We conducted 20 runs in each setting. Figure 7.2 shows the success rates that were observed. We note that the results appear to indicate that the relaxation provided by truncation damages problem-solving performance, with greater relaxation generally producing greater damage. We have too little data to draw any firm conclusions about the extent of this effect, or about the ways in which this effect may vary across problems, but we present it as a baseline for future study of truncated lexicase selection.

7.8   Relation to Many-Objective Optimization

All of the lexicase selection relaxation algorithms discussed above assess programs with respect to multiple test cases, each of which might be considered, in some sense, to be a separate objective of the genetic programming search process. Because there is existing research on so-called "many-objective optimization algorithms," which attempt to optimize four or more objectives, it is worth considering how such research relates to the parent selection algorithms we have discussed here.

We recently provided a multi-objective analysis of lexicase selection to show that, if we treat each fitness case as an objective, parents selected by lexicase selection are Pareto-optimal (i.e., they are non-dominated in the population) and located at the boundaries of the Pareto front [11]; we borrow from that analysis in the remainder of this section. The utility of Pareto dominance as a search driver is reduced as the number of fitness cases/objectives grows since the number of non-dominated solutions grow exponentially with the number of objectives. However, it is noteworthy that lexicase selection corresponds to selections of the boundaries of the Pareto front since boundary solutions influence typical measures of quality in many-objective optimization.

In many-objective optimization, the performance of algorithms is typically assessed in terms of convergence, uniformity, and spread [13], with the last of these dealing directly with the extent of boundary solutions. Some indicator-based methods, for example IBEA and SMS-EMOA, use a measure of the hypervolume in objective space to evaluate algorithm performance [23]. There appears to be some disagreement regarding the importance of boundary solutions in this context. Although Deb et al. [2] argued empirically that boundary solutions have an outsized effect on hypervolume measures, according to Auger et al., "optimizing the unweighted hypervolume indicator stresses the so-called knee-points—parts of the Pareto front decision-makers believe to be interesting regions...Extreme points are not generally preferred as claimed in [2], since the density of points does not depend on the position on the front but only on the gradient at the respective point" [1].

In general, many-objective optimization methods have not highlighted random sampling as a useful method for exploring high-dimensional objective spaces. MOEA/D, R-NSGA-II, and NSGA-III opt for the use of reference points in objective space to preserve the spread of solutions. Unlike lexicase selection, there does not seem to be explicit motivation to keep boundary solutions in this literature.

Results like those demonstrated in the present chapter, and in other work on lexicase selection, suggest that it may be worthwhile for many-objective optimization researchers to give greater consideration to methods based on random sampling, and to methods in which samples are considered with randomized priorities.

7.9   Discussion

In ordinary lexicase selection, test case order is randomized, allowing the algorithm to implicitly select on the basis of performance on all subsets of test cases, even though the exponentially many subsets are not considered explicitly. Randomization samples the space of test case sequence prefixes, thereby sampling the space of test case combinations that "matter" for a particular parent selection event, and thereby implicitly considering, over time, all subsets of test cases.

This appears to be a powerful technique for selecting useful parents in genetic programming, as it sometimes allows solutions to be found in significantly more runs and/or fewer generations than they can be found with previous, test-case-aggregating parent selection algorithms. However, the ordinary lexicase selection algorithm appears to be too strict in some circumstances, preventing the selection of parents that have much to offer to future generations in the evolutionary search process.

Epsilon lexicase selection appears to resolve this issue for many problems with floating-point errors by relaxing the requirement for eliteness in each step of the filtering of candidate parents. Two closely related methods were also explored here. The first, random threshold lexicase selection, is often (but not necessarily always) a more (and randomly more) relaxed version of epsilon lexicase selection, which also depends less directly on the distribution of errors in the population than does epsilon lexicase selection. The second, MADCAP epsilon lexicase selection, is a less (and randomly less) relaxed version of epsilon lexicase selection. Neither of these alternatives performed better than epsilon lexicase selection on the problems studied here, but they both performed better than several of the other methods considered, and one can imagine situations in which each would be more useful. For example, one might expect random threshold lexicase selection to be useful in contexts requiring particularly broad exploration, and one might expect MADCAP epsilon lexicase selection to be useful in contexts requiring convergence to solutions with zero or truly minimal errors. One area for future research is the application of these methods to problems of other types, in order to support or to falsify such expectations.

Truncated lexicase selection provides a different form of relaxation, limiting the number of test cases that can be considered in any parent selection event. While we were initially motivated to perform experiments on truncated lexicase selection by an expectation that the truncation might provide runtime performance benefits, we also wanted to study how this affects solution rates. Our experiments here, while preliminary, suggest that truncation often hampers the ability of the system to find solutions, but that it does so to a different extent on different problems. These results suggest that more experiments should be conducted, on additional problems of various types, to help us to understand when this form of relaxation might be beneficial (regarding success rates and/or runtime) and when it is detrimental. They also suggest that experiments should be conducted with additional truncation levels. Levels of 50% or higher may be particularly interesting, as they would provide only modest relaxation relative to ordinary lexicase selection. Levels that allow the consideration of very few test cases, but more than just a single one, may also be revealing because they would be quite relaxed while still allowing for selection on the basis of multiple test cases.

Another obvious avenue for future research is to combine truncation with relaxation of the eliteness constraints for individual test cases. That is, it would be straightforward and possibly useful or at least instructive to conduct runs with truncated versions of epsilon lexicase selection, random threshold lexicase selection, and MADCAP epsilon lexicase selection.

Other forms of relaxation are also possible and may be useful in some circumstances. Two that we have implemented but not yet studied systematically are methods in which we perform lexicase selection with a certain probability and purely random selection otherwise, and methods in which we perform truncated lexicase selection but with different, randomly chosen numbers of test cases used in each selection event. Of course, these could also be combined with one another, and with many of the other methods described above. The choice of which of these to explore first might best be guided by theoretical consideration of the ways in which they sample the search space, along the lines of the initial discussion that we offered above on the relation between work on lexicase selection and work on many-objective optimization.

We note that many of the ideas discussed here could be applied to survival selection as well as parent selection.

While all of the methods here in some way relax the requirements made by ordinary lexicase selection, it is possible that strengthening the requirements in some way could potentially have benefits for some problems. It is unclear at this point what such strengthening would look like, but variants of this sort would certainly be interesting to examine.

The specific methods described here appear to have varying utility, at least from the experiments conducted to date. Regardless of the utility of the specific methods, however, we hope that the discussion here may help to stimulate additional work developing selection algorithms that can guide evolution more effectively.

Acknowledgements This material is based upon work supported by the National Science Foundation under Grants No. 1617087, 1129139 and 1331283. Any opinions, findings, and conclusions or recommendations expressed in this publication are those of the authors and do not necessarily reflect the views of the National Science Foundation.

References

1. Anne Auger, Johannes Bader, Dimo Brockhoff, and Eckart Zitzler. Theory of the hypervolume indicator: optimal -distributions and the choice of the reference point. In *Proceedings of the tenth ACM SIGEVO workshop on Foundations of genetic algorithms*, pages 87–102. ACM, 2009.
2. Kalyanmoy Deb, Manikanth Mohan, and Shikhar Mishra. Evaluating the ε-Domination Based Multi-Objective Evolutionary Algorithm for a Quick Computation of Pareto-Optimal Solutions. *Evolutionary Computation*, 13(4):501–525, December 2005.
3. Thomas Helmuth, Nicholas Freitag McPhee, and Lee Spector. Effects of lexicase and tournament selection on diversity recovery and maintenance. In Tobias Friedrich and et al., editors, *GECCO '16 Companion: Proceedings of the Companion Publication of the 2016 Annual Conference on Genetic and Evolutionary Computation*, pages 983–990, Denver, Colorado, USA, 20–24 July 2016. ACM.
4. Thomas Helmuth, Nicholas Freitag McPhee, and Lee Spector. The impact of hyperselection on lexicase selection. In Tobias Friedrich, editor, *GECCO '16: Proceedings of the 2016 Annual Conference on Genetic and Evolutionary Computation*, pages 717–724, Denver, USA, 20–24 July 2016. ACM. Nominated for best paper.
5. Thomas Helmuth, Nicholas Freitag McPhee, and Lee Spector. Lexicase selection for program synthesis: A diversity analysis. In Rick Riolo, William P. Worzel, M. Kotanchek, and A. Kordon, editors, *Genetic Programming Theory and Practice XIII*, Genetic and Evolutionary Computation, pages 151–167, Ann Arbor, USA, May 2016. Springer.
6. Thomas Helmuth and Lee Spector. General program synthesis benchmark suite. In Sara Silva and et al., editors, *GECCO '15: Proceedings of the 2015 Annual Conference on Genetic and Evolutionary Computation*, pages 1039–1046, Madrid, Spain, 11–15 July 2015. ACM.
7. Thomas Helmuth, Lee Spector, and James Matheson. Solving uncompromising problems with lexicase selection. *IEEE Transactions on Evolutionary Computation*, 19(5):630–643, October 2015.
8. Krzysztof Krawiec and Una-May O'Reilly. Behavioral programming: A broader and more detailed take on semantic gp. In *Proceedings of the 2014 Annual Conference on Genetic and Evolutionary Computation*, GECCO '14, pages 935–942, New York, NY, USA, 2014. ACM.
9. Krzysztof Krawiec, Jerry Swan, and Una-May O'Reilly. Behavioral program synthesis: Insights and prospects. In *Genetic Programming Theory and Practice XIII*, Genetic and Evolutionary Computation. Springer, 2015.
10. William La Cava, Kourosh Danai, and Lee Spector. Inference of compact nonlinear dynamic models by epigenetic local search. *Engineering Applications of Artificial Intelligence*, 55:292–306, October 2016.
11. William La Cava, Thomas Helmuth, Lee Spector, and Jason H. Moore. ε-Lexicase selection: a probabilistic and multi-objective analysis of lexicase selection in continuous domains. *Evolutionary Computation*, 1–28. https://doi.org/10.1162/evco_a_00224.
12. William La Cava, Lee Spector, and Kourosh Danai. Epsilon-lexicase selection for regression. In Tobias Friedrich, editor, *GECCO '16: Proceedings of the 2016 Annual Conference on Genetic and Evolutionary Computation*, pages 741–748, Denver, USA, 20–24 July 2016. ACM.
13. Miqing Li and Jinhua Zheng. Spread assessment for evolutionary multi-objective optimization. In *International Conference on Evolutionary Multi-Criterion Optimization*, pages 216–230. Springer, 2009.
14. M. Lichman. UCI machine learning repository, 2013.
15. Pawel Liskowski, Krzysztof Krawiec, Thomas Helmuth, and Lee Spector. Comparison of semantic-aware selection methods in genetic programming. In Colin Johnson, Krzysztof Krawiec, Alberto Moraglio, and Michael O'Neill, editors, *GECCO 2015 Semantic Methods in Genetic Programming (SMGP'15) Workshop*, pages 1301–1307, Madrid, Spain, 11–15 July 2015. ACM.

16. Samir W Mahfoud. *Niching methods for genetic algorithms*. PhD thesis, 1995.
17. Yuliana Martnez, Enrique Naredo, Leonardo Trujillo, Pierrick Legrand, and Uriel Lpez. A comparison of fitness-case sampling methods for genetic programming. *Journal of Experimental & Theoretical Artificial Intelligence*, 29(6):1203–1224, 2017.
18. Nicholas Freitag McPhee, Brian Ohs, and Tyler Hutchison. Semantic building blocks in genetic programming. In *Proceedings of the 11th European Conference on Genetic Programming, EuroGP 2008*, volume 4971 of *Lecture Notes in Computer Science*, pages 134–145, Naples, 26–28 March 2008. Springer.
19. Michael Schmidt and Hod Lipson. Age-fitness Pareto optimization. In *Genetic Programming Theory and Practice VIII*, pages 129–146. Springer, 2011.
20. Lee Spector. Assessment of problem modality by differential performance of lexicase selection in genetic programming: A preliminary report. In Kent McClymont and Ed Keedwell, editors, *1st workshop on Understanding Problems (GECCO-UP)*, pages 401–408, Philadelphia, Pennsylvania, USA, 7–11 July 2012. ACM.
21. Robert Tibshirani. Regression shrinkage and selection via the lasso. *Journal of the Royal Statistical Society. Series B (Methodological)*, pages 267–288, 1996.
22. Sarah Anne Troise and Thomas Helmuth. Lexicase selection with weighted shuffle. In *Genetic Programming Theory and Practice XV*, Genetic and Evolutionary Computation, pages 89–103, Ann Arbor, USA, May 2017. Springer.
23. Tobias Wagner, Nicola Beume, and Boris Naujoks. Pareto-, Aggregation-, and Indicator-Based Methods in Many-Objective Optimization. In *Evolutionary Multi-Criterion Optimization*, pages 742–756. Springer, Berlin, Heidelberg, March 2007. https://doi.org/10.1007/978-3-540-70928-2_56.

Chapter 8
A System for Accessible Artificial Intelligence

Randal S. Olson, Moshe Sipper, William La Cava, Sharon Tartarone,
Steven Vitale, Weixuan Fu, Patryk Orzechowski, Ryan J. Urbanowicz,
John H. Holmes, and Jason H. Moore

Abstract While artificial intelligence (AI) has become widespread, many
commercial AI systems are not yet accessible to individual researchers nor the
general public due to the deep knowledge of the systems required to use them. We
believe that AI has matured to the point where it should be an accessible technology
for everyone. We present an ongoing project whose ultimate goal is to deliver
an open source, user-friendly AI system that is specialized for machine learning
analysis of complex data in the biomedical and health care domains. We discuss
how genetic programming can aid in this endeavor, and highlight specific examples
where genetic programming has automated machine learning analyses in previous
projects.

R. S. Olson · W. La Cava · S. Tartarone · S. Vitale · W. Fu · R. J. Urbanowicz · J. H. Holmes
J. H. Moore (✉)
Institute for Biomedical Informatics, University of Pennsylvania, Philadelphia, PA, USA
e-mail: rso@randalolson.com; lacava@upenn.edu; ryanurb@pennmedicine.upenn.edu;
jhmoore@upenn.edu

M. Sipper
Institute for Biomedical Informatics, University of Pennsylvania, Philadelphia, PA, USA

Department of Computer Science, Ben-Gurion University, Beer-Sheva, Israel
e-mail: sipper@cs.bgu.ac.il

P. Orzechowski
Institute for Biomedical Informatics, University of Pennsylvania, Philadelphia, PA, USA

Department of Automatics and Biomedical Engineering, AGH University of Science
and Technology, Krakow, Poland

© Springer International Publishing AG, part of Springer Nature 2018
W. Banzhaf et al. (eds.), *Genetic Programming Theory and Practice XV*,
Genetic and Evolutionary Computation, https://doi.org/10.1007/978-3-319-90512-9_8

8.1 Introduction

A central goal of artificial intelligence (AI) is to use computational hardware and software to solve complex problems in a human-competitive manner [9]. The practicality of this goal is that AI can be tasked with solving problems or performing functions that humans cannot perform or simply do not have time for. Most AI methodologies can be grouped into top-down approaches, wherein cognition is viewed as a high-level phenomenon that is independent of the low-level details, or bottom-up approaches, which define basic computational building blocks such as artificial neurons that collectively give rise to "emergent" [29] intelligent behavior. The top-down approach has been difficult to realize given the inherent complexity of human cognition. However, the bottom-up has had some success owing to the availability of sophisticated algorithms such as genetic programming (GP) [10] and deep neural networks [6]. This is particularly true today with abundant and inexpensive high-performance computing, leading to many human-competitive success stories [9].

Medical applications of AI have had a long history with both successes and failures. One of the early successes was the Mycin system, which was designed to predict the antibiotic that a patient with an infection should receive in the intensive care unit [2]. Mycin combined a knowledge base along with a set of rules implemented as part of an expert system. The system was demonstrated to be human-competitive, but was never put into clinical practice because of legal concerns and the time it took clinicians to enter the patient data required for Mycin to make the predictions. The field of AI has matured since Mycin was developed and, importantly, computing power has grown tremendously in parallel. Examples of modern AI successes include IBM's Watson, which beat the world champion of the game show Jeopardy [5]. The Watson AI system that won Jeopardy combined knowledge representation, information retrieval, natural language processing, and machine learning along with high-performance computing to access and exploit a knowledge base that included the Wikipedia text corpus. This was a milestone in AI because it showed that a computational system could compete with humans on difficult language processing tasks. Watson is now being marketed in the health care domain although the jury is still out on its effectiveness.

Commercial AI systems such as Watson show potential but are not yet accessible to individual researchers nor the general public due the cost and the complexity of working with a team from IBM. It is our working hypothesis that,

AI has matured to the point where it should be an accessible technology for everyone.

Democratization of AI will be important if we seek to integrate this exciting new technology into multiple different domains, as demonstrated by recent efforts such as Orange [4]. We describe here the early development stages of an open source and user-friendly AI system—PennAI (http://pennai.org)—for machine learning analysis of complex data in the biomedical and health care domains. We focus our initial efforts on the classification of biomedical endpoints such as disease susceptibility. We describe in turn below each of the components of our AI system and then end with an example and a discussion of how we envision this system

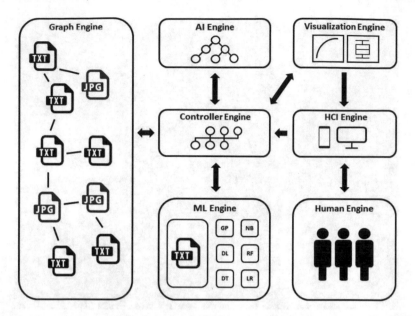

Fig. 8.1 The components of PennAI, a user-friendly AI system developed at the University of Pennsylvania

being used to solve complex biomedical problems. Further, we discuss how GP can aid in enhancing PennAI, and highlight specific examples where GP has automated machine learning analyses in previous work.

The components of PennAI include a human engine (i.e., the user); a user-friendly interface for interacting with the AI; a machine learning engine for data mining; a controller engine for launching jobs and keeping track of analytical results; a graph database for storing data and results (i.e., the memory); an AI engine for monitoring results and automatically launching or recommending new analyses; and a visualization engine for displaying results and analytical knowledge (Fig. 8.1). This AI system provides a comprehensive set of integrated components for automated machine learning (AutoML), thus providing a data science assistant for generating useful results from large and complex data problems. PennAI is housed in the "Idea Factory," a facility designed to facilitate collaboration and promote new methods of communicating and presenting scientific innovation. The Idea Factory makes sophisticated data visualization and artificial intelligence analytics easy for users across the entire Penn community (Fig. 8.2).

8.2   The Human Engine

The most important component of the proposed AI system is the user. Contrary to some claims that AI will replace human users, we see the human as an integral part of the discovery process and a partner with the AI. One way to view this

Fig. 8.2 The "Idea Factory," home of PennAI

partnership is with the human as the driver of the discovery process and the AI as the data science assistant. Thus, the AI provides an additional set of hands in a modern data science discovery environment that might include human teammates with expertise in computer science, statistics, and applied mathematics. We have previously suggested this idea of human-computer interaction that places the human user at the epicenter [22]. This idea has also previously been explored from the point of view of the user or domain expert [16].

Langley [16] provides five important tips that are relevant to thinking about the relationship humans have with AI for data mining using machine learning. First, traditional machine learning notations are not easily communicated to scientists. This consideration is important because a machine learning model may not be interpretable by a user. Second, scientists often have initial models that should influence the discovery process. Domain-specific knowledge can be critical to the discovery process. Third, scientific datasets are often rare and difficult to obtain. It often takes years to collect and process the data before it can be analyzed. As such, it is important that the analysis is carefully planned and executed, and that any general feedback about the performance of the learning process is not lost between studies. Fourth, scientists want models that move beyond description and provide explanations of the data. Explanation and interpretation are paramount to the user. Finally, scientists want computational assistance rather than a complete replacement of themselves. Langley [16] further suggests that users want interactive discovery environments that help them understand their data while at the same time giving them control over the modeling process. Collectively, these five lessons suggest that synergy between the user and the AI is critical. With this in mind, our proposed AI system includes a graphical user interface (GUI) that allows the user to easily launch analyses, view the results, and give the AI feedback about what results are useful or interesting.

8.3   The Human-Computer Interaction Engine

As described above, a key component of PennAI is human-computer interaction. The first important feature is to make it easy for the user to directly launch machine learning analyses by choosing a method and its parameter settings from an intuitive push-button menu implemented through the web using JavaScript. The user can launch single analyses or, in an advanced mode, launch a grid search across multiple methods and parameter settings. The methods and the controller that keep track of these analyses is described below. Figures 8.3 and 8.4 show prototypes of our GUI for uploading and viewing datasets for analysis and launching machine learning analyses on those datasets, respectively. Our JavaScript implementation is compatible with mobile devices, which allows the user to interact with the AI system from any Internet-connected device.

The second key feature of PennAI is the ability to toggle the AI on and off for automated analysis, shown in Fig. 8.3. An AI toggle allows the user to turn the AI on and set parameters controlling the maximum number of runs the AI can launch, as well as the frequency of updates the user would like to receive by email or text message. The GUI also provides a simple thumbs up/down selection for each result received by PennAI, which provides feedback to PennAI that is incorporated into its expert knowledge system.

8.4   The Machine Learning Engine

Our first application of PennAI is for data mining using machine learning in the biomedical domain. Here, we make use of an extensive open source machine learning library in Python called scikit-learn [28]. Scikit-learn provides peer-reviewed implementations of several common supervised and unsupervised machine learning algorithms, data preprocessing methods, feature engineering and selection methods, hyperparameter optimization procedures, and more. To most users, scikit-learn is considered to be the standard machine learning library in Python.

Of course, there are dozens of machine learning algorithms, preprocessors, etc. to choose from in scikit-learn, which can be overwhelming to a novice user. To simplify the algorithm selection process for PennAI users, we currently limit PennAI to six machine learning algorithms that we believe will handle most supervised classification use cases, shown in Table 8.1. We also limit the parameter choices for each algorithm to a handful of the most important parameters and parameter

Fig. 8.3 Prototype of the graphical user interface for managing and viewing datasets

Table 8.1 Machine learning algorithms available in PennAI

Classification	Regression
Logistic regression	ElasticNet
Decision tree	Decision tree
k-Nearest neighbors	k-Nearest neighbors
Support vector machine	Support vector machine
Random forest	Random forest
Gradient boosting	Gradient boosting

options, which makes it easier for users to choose a parameter configuration at the expense of algorithm customizability. An example of the interface to the Machine Learning Engine can be found in Fig. 8.4, where only a handful of the most important parameters and parameter options are available for the k-Nearest Neighbors classification algorithm.

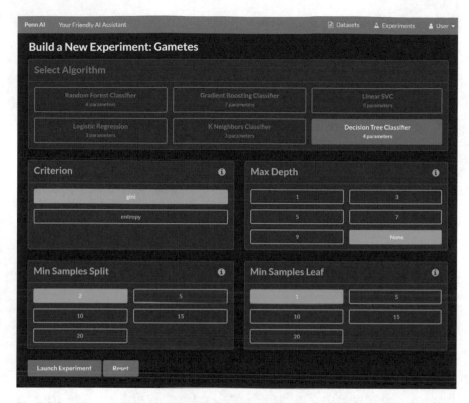

Fig. 8.4 Prototype of the Machine Learning Engine graphical user interface

In an upcoming PennAI implementation, we will provide simplified descriptions
of the machine learning algorithms and parameters so users can make use of the
algorithms without fully understanding their implementation. For example, when
using a random forest it is not necessary for the user to understand what tuning the
n_estimators parameter does to the model. Instead, it is more important for
the user to understand that adding more decision trees to the random forest (i.e.,
increasing n_estimators) improves model performance but increases training
time, whereas removing decision trees from the random forest decreases model
performance but decreases training time [7].

Once the Machine Learning Engine finishes training and evaluating a machine
learning model, it stores the machine learning model, the model predictions, and
an analysis of the model in the Graph Database Engine, which are used in the
Visualization Engine (both described below).

8.5   The Controller Engine

The Controller Engine acts as the interface between the high-performance comput-ing system and the user or AI. This component is hidden from the user but facilitates the automatic launching of jobs on a multi-CPU machine, computing cluster, or cloud computing system. The controller must not only coordinate the launching of jobs but also keep track of when they finish and deposit the results in the Graph Database Engine (described below) that serves as the memory of the system.

For the Controller Engine, we selected an open source package called the Future Gadget Lab (FGLab), which is available on GitHub.[1] FGLab functions as a server with individual runs launched as clients, called FGMachines. FGLab uses node.js to coordinate distributed jobs and uses MongoDB [3] as the backend database in the Graph Database Engine.

8.6   The Graph Database Engine

Another key component of PennAI is a memory system that keeps track of every analysis that is run on each data set. We keep track of the details of the machine learning method, the parameter settings, the data set analyzed, and results such as the model, model error, and area under the receiver operating characteristic curve (AUC). These are all stored in a JSON file that is deposited in a MongoDB NoSQL database. The advantage of using a NoSQL database is that new data elements can be added without creating tables and without strict format specifications. This flexibility is important for the rapidly changing landscape of machine learning. MongoDB can also function as a graph database that allows the documents to be linked in a network according to shared index terms related to the analysis and data. This feature facilitates more complex semantic queries of the database, such as "Return the machine learning algorithm configurations that achieved the highest accuracy on any study involving prostate cancer."

8.6.1   Knowledge Base

The Graph Database Engine serves as the memory of PennAI and provides the raw materials for the AI to learn which methods and parameter settings are working better than others for particular kinds of problems. The initial knowledge base consists of results from a previously published benchmark of scikit-learn algorithms [24], in which 14 machine learning algorithms were run with full

---

[1]FGLab: https://github.com/Kaixhin/FGLab.

hyperparameter optimization on a suite of 165 supervised classification problems. The results are combined with meta-information about the datasets (e.g., number of features, number of instances, correlations between features, etc.) in order to allow the creation of a mapping from 'problem instance space', i.e. dataset meta-features and model performance, to 'learning space', i.e. machine learning algorithms and their parameters. This data can then be modelled to extract rules that represent the knowledge used by the Artificial Intelligence Engine to make informed analyses. The knowledge base will be updated with all future analyses.

8.7   The Artificial Intelligence Engine

Each component described above provides the raw materials for the Artificial Intelligence Engine which then (1) searches the graph database for results related to one or more data sets, (2) performs statistical analysis comparing algorithms and their parameters, (3) combines facts and rules in an expert system to make new analysis recommendations, (4) communicates findings to the user, and (5) automatically launches new analyses using suggestions from the expert system. The first function uses the search capabilities of the MongoDB graph database to identify relevant machine learning results in the form of JSON files. All returned JSON files can be parsed to extract the machine learning algorithm, parameters, and information about the model performance. These results are collated in a tab-delimited file and a statistical analysis performed to determine the best algorithm configurations for certain problem types, similar to meta-learning techniques [8].

New statistical results are used to populate the knowledge base of an expert system that has a set of decision rules provided by developers and advanced machine learning practitioners. This expert system is then used to make suggestions for additional analyses, for example by recommending better parameter settings or even entirely different machine learning algorithms that might be better-suited for the user's dataset. The user can access these suggestions manually or PennAI can use the suggestions to automatically launch new jobs, thus continually growing the PennAI knowledge base. Essentially, the Artificial Intelligence Engine becomes a research assistant who tinkers with new ways of modeling the dataset and reports back to the user with their best findings.

8.8   The Visualization Engine

Visualization will be critical for fostering the human-AI collaboration described above. The user will need to be able to see individual machine learning models and results as well as higher-level results from statistical analyses across machine learning runs. We extract visual results such as the receiver operating characteristic (ROC) curves and models to store in the graph database, as shown in Fig. 8.5.

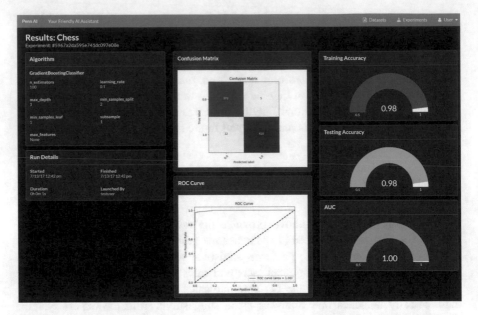

Fig. 8.5 Prototype of the Visualization Engine graphical user interface

PennAI will also generate heatmaps and other visualizations that summarize results across different machine learning methods and datasets. These higher-level visualizations will aid the user with making decisions about new manual analyses to launch and will help them assess how well the PennAI assistant is doing. These images will be linked to the datasets and results in the Graph Database Engine, and will thus be easily searchable.

8.9   Discussion and Future Work

Thus far, we have described PennAI as a system that provides a simple interface for users to upload their datasets, launch machine learning analyses, view the results of the analyses in an intuitive manner, and use those results to refine their machine learning analyses. We also described how PennAI will use a combination of expert knowledge from advanced machine learning practitioners and prior statistical knowledge of machine learning algorithm performance on datasets to recommend new analyses to the user, as well as launch its own analyses to later report to the user. In essence, the primary goal of PennAI is to provide an AI research assistant for its users. However, considering the name of this workshop and book—Genetic Programming Theory and Practice—one may be left wondering how GP can be incorporated into PennAI. In the following paragraphs, we will describe our plans for integrating GP into PennAI.

The first point of entry is to include GP as a machine learning option since a number of successful biomedical applications have been reported (e.g., [17–21, 34]). A GP system for classification based on multidimensional clustering [31] was recently demonstrated on biomedical classification problems [15] as a competitive alternative to traditional machine learning approaches. Recently GP has been proposed as a general feature engineering wrapper (FEW)[2] in order to harness its feature learning capability to improve scikit-learn estimators, both for regression [13] and classification [14]. FEW allows GP to provide readable feature transformations to users while still utilizing existing modeling techniques for making predictions. As mentioned in Sect. 8.2, interpretation and explanatory power are important aspects of using AI for data mining, and therefore GP methods that produce concise models, e.g. by local search [11] or Pareto optimization [12], are important options to include. Further down the road, it could be possible for PennAI to allow advanced users to incorporate custom machine learning algorithms into PennAI by providing a scikit-learn formatted interface to their project (e.g. ellyn[3]). PennAI could then provide a "bring your own learner" type of service [1] to allow researchers to tackle complex data mining tasks with customized learning approaches, and incorporate the results into its knowledge base for improving future data science projects.

Beyond using GP to perform the machine learning itself, recent work has shown that GP can also be harnessed to optimize a sequence of existing data analysis and machine learning operations on a dataset to maximize the predictive performance of the final machine learning model [30, 35]. For example, TPOT[4] is an early prototype that uses GP to optimize a sequence of scikit-learn operations for both classification and regression problems [25–27], and has been shown to work quite well across a broad range of application domains ranging from epidemiological studies to image classification to time series prediction [23]. Given the general design of TPOT, the operations it optimizes over can be specialized for particular problem domains. As another example, the TPOT-MDR project [33] showed that TPOT can be specialized for genome-wide association studies (GWAS), and it outperforms several state-of-the-art modeling methods on both simulated and real-world GWAS problems because it considers a broad range of operations in with one another. As such, we view GP as a strong candidate for a future version of the PennAI Artificial Intelligence Engine, where the GP is seeded with the best known algorithm configurations and uses the core principles of GP (inheritance, mutation, and crossover)—distributed over a high-performance computing cluster— to improve the algorithm configurations from there. This brand of GP-based AI system would be useful for automatically launching new analyses, but less useful for recommending particular algorithm configurations to the user because GP does not provide a notion of the "next best" solution to attempt.

---

[2]http://lacava.github.io/few.

[3]http://epistasislab.github.io/ellyn.

[4]https://github.com/rhiever/tpot.

Another extension of PennAI is the use of a meta genetic algorithm to find parameters (population size, generation count, etc.) for a GP instance that work well, i.e., solve a given problem [32]. This meshes well with the idea that the AI of PennAI will aid non-machine learning experts run complex algorithms, such as GP, without having to find or even understand every single parameter.

Ultimately, PennAI will likely be comprised of several disparate AI algorithms that use meta-data and meta-learning to improve the user experience and user productivity by suggesting machine learning algorithms and parameters, as well as providing other insights. As a result, we will be able to harness ensemble techniques to collate the advice given by the numerous AI algorithms.

The time is now to bring AI technology to anyone that wants to use it for big data analytics. The software and hardware technology exists and data has never been bigger, more complex, and more plentiful. PennAI will provide both machine learning and AI capability to both naive and expert users alike with a user-friendly web and smartphone-enabled interface. We see AI technology such as PennAI not as a replacement for the data scientist but rather as a data science assistant that can suggest analyses to the user or provide automatically generated results that are informed by previous analyses across different data sets. The user can take these results as-is or use them as inspiration in manual analyses. The democratization of AI is here.

Acknowledgements This work was generously funded by the Perelman School of Medicine and the University of Pennsylvania Health System. Additional funding was provided by National Institutes of Health grants AI116794, DK112217, ES013508, and TR001878.

References

1. Arnaldo, I., Veeramachaneni, K., Song, A., O'Reilly, U.M.: Bring your own learner: A cloud-based, data-parallel commons for machine learning. IEEE Computational Intelligence Magazine **10**(1), 20–32 (2015)
2. Bruce, G., Buchanan, B., Shortliffe, E.: Rule-based expert systems: The MYCIN experiments of the Stanford heuristic programming project (1984)
3. Chodorow, K., Dirolf, M.: MongoDB: The Definitive Guide, 1st edn. O'Reilly Media, Inc. (2010)
4. Demšar, J., Curk, T., Erjavec, A., Črt Gorup, Hočevar, T., Milutinovič, M., Možina, M., Polajnar, M., Toplak, M., Starič, A., Štajdohar, M., Umek, L., Žagar, L., Žbontar, J., Žitnik, M., Zupan, B.: Orange: Data mining toolbox in Python. Journal of Machine Learning Research **14**, 2349–2353 (2013)
5. Ferrucci, D.A.: Introduction to "This is Watson". IBM Journal of Research and Development **56**(3.4), 1–1 (2012)
6. Goodfellow, I., Bengio, Y., Courville, A.: Deep learning. MIT Press (2016)
7. Hastie, T.J., Tibshirani, R.J., Friedman, J.H.: The Elements of Statistical Learning: Data Mining, Inference, and Prediction. Springer, New York, NY, USA (2009)
8. Kalousis, A.: Algorithm selection via meta-learning. Ph.D. thesis, Universite de Geneve (2002)
9. Kannappan, K., Spector, L., Sipper, M., Helmuth, T., La Cava, W., Wisdom, J., Bernstein, O.: Analyzing a decade of human-competitive ("HUMIE") winners: What can we learn? In: Genetic Programming Theory and Practice XII, pp. 149–166. Springer International Publishing (2015)

10. Koza, J.R.: Genetic programming: on the programming of computers by means of natural selection, vol. 1. MIT Press (1992)
11. La Cava, W., Danai, K., Spector, L.: Inference of compact nonlinear dynamic models by epigenetic local search. Engineering Applications of Artificial Intelligence **55**, 292–306 (2016)
12. La Cava, W., Danai, K., Spector, L., Fleming, P., Wright, A., Lackner, M.: Automatic identification of wind turbine models using evolutionary multiobjective optimization. Renewable Energy **87**, 892–902 (2016)
13. La Cava, W., Moore, J.: A general feature engineering wrapper for machine learning using ϵ-lexicase survival. In: European Conference on Genetic Programming, pp. 80–95. Springer (2017)
14. La Cava, W., Moore, J.H.: Ensemble representation learning: an analysis of fitness and survival for wrapper-based genetic programming methods. In: GECCO '17: Proceedings of the Conference on Genetic and Evolutionary Computation. ACM (2017)
15. La Cava, W., Silva, S., Vanneschi, L., Spector, L., Moore, J.: Genetic programming representations for multi-dimensional feature learning in biomedical classification. In: European Conference on the Applications of Evolutionary Computation, pp. 158–173. Springer (2017)
16. Langley, P.: Lessons for the Computational Discovery of Scientific Knowledge (2002)
17. Moore, J.H., Andrews, P.C., Barney, N., White, B.C.: Development and evaluation of an open-ended computational evolution system for the genetic analysis of susceptibility to common human diseases. In: European Conference on Evolutionary Computation, Machine Learning and Data Mining in Bioinformatics, pp. 129–140. Springer (2008)
18. Moore, J.H., Greene, C.S., Hill, D.P.: Identification of novel genetic models of glaucoma using the "emergent" genetic programming-based artificial intelligence system. In: R. Riolo, W.P. Worzel, M. Kotanchek (eds.) Genetic Programming Theory and Practice XII, pp. 17–35. Springer International Publishing, Cham (2015)
19. Moore, J.H., Greene, C.S., Hill, D.P.: Identification of novel genetic models of glaucoma using the "emergent" genetic programming-based artificial intelligence system. In: Genetic Programming Theory and Practice XII, pp. 17–35. Springer (2015)
20. Moore, J.H., Hill, D.P., Fisher, J.M., Lavender, N., Kidd, L.C.: Human-computer interaction in a computational evolution system for the genetic analysis of cancer. In: R. Riolo, E. Vladislavleva, J.H. Moore (eds.) Genetic Programming Theory and Practice IX, pp. 153–171. Springer New York, New York, NY (2011)
21. Moore, J.H., Hill, D.P., Saykin, A., Shen, L.: Exploring interestingness in a computational evolution system for the genome-wide genetic analysis of alzheimer's disease. In: R. Riolo, J.H. Moore, M. Kotanchek (eds.) Genetic Programming Theory and Practice XI, pp. 31–45. Springer New York, New York, NY (2014)
22. Moore, J.H., White, B.C.: Genome-wide genetic analysis using genetic programming: The critical need for expert knowledge. In: Genetic Programming Theory and Practice IV, pp. 11–28. Springer (2007)
23. Olson, R.S., Bartley, N., Urbanowicz, R.J., Moore, J.H.: Evaluation of a Tree-based Pipeline Optimization Tool for Automating Data Science. In: GECCO 2016, GECCO '16, pp. 485–492. ACM, New York, NY, USA (2016)
24. Olson, R.S., La Cava, W., Orzeshowski, P., Urbanowicz Ryan J Moore, J.H.: PMLB: A large benchmark suite for machine learning evaluation and comparison. arXiv e-print. https://arxiv.org/abs/1703.00512 (2017)
25. Olson, R.S., Moore, J.H.: Identifying and Harnessing the Building Blocks of Machine Learning Pipelines for Sensible Initialization of a Data Science Automation Tool. arXiv e-print. http://arxiv.org/abs/1607.08878 (2016)
26. Olson, R.S., Moore, J.H.: TPOT: A Tree-based Pipeline Optimization Tool for Automating Machine Learning. JMLR **64**, 66–74 (2016)
27. Olson, R.S., Urbanowicz, R.J., Andrews, P.C., Lavender, N.A., Kidd, L.C., Moore, J.H.: Automating Biomedical Data Science Through Tree-Based Pipeline Optimization. In: G. Squillero, P. Burelli (eds.) Applications of Evolutionary Computation: 19th European Conference, EvoApplications 2016, Porto, Portugal, March 30–April 1, 2016, Proceedings, Part I, pp. 123–137. Springer International Publishing (2016)

28. Pedregosa, F., Varoquaux, G., Gramfort, A., Michel, V., Thirion, B., Grisel, O., Blondel, M., Prettenhofer, P., Weiss, R., Dubourg, V., et al.: Scikit-learn: Machine learning in Python. Journal of Machine Learning Research 12, 2825–2830 (2011)
29. Ronald, E.M., Sipper, M., Capcarrère, M.S.: Design, observation, surprise! A test of emergence. Artificial Life 5(3), 225–239 (1999)
30. de Sá, A.G., Pinto, W.J.G., Oliveira, L.O.V., Pappa, G.L.: RECIPE: A Grammar-Based Framework for Automatically Evolving Classification Pipelines. In: European Conference on Genetic Programming, pp. 246–261. Springer (2017)
31. Silva, S., Muñoz, L., Trujillo, L., Ingalalli, V., Castelli, M., Vanneschi, L.: Multiclass classification through multidimensional clustering. In: Genetic Programming Theory and Practice XIII, pp. 219–239. Springer (2016)
32. Sipper, M., Fu, W., Ahuja, K., Moore, J.H.: Investigating the parameter space of evolutionary algorithms (2017). arXiv:1706.04119
33. Sohn, A., Olson, R.S., Moore, J.H.: Toward the automated analysis of complex diseases in genome-wide association studies using genetic programming. In: Proceedings of the Genetic and Evolutionary Computation Conference, GECCO '17, pp. 489–496. ACM, New York, NY, USA (2017)
34. Vanneschi, L., Archetti, F., Castelli, M., Giordani, I.: Classification of oncologic data with genetic programming. Journal of Artificial Evolution and Applications p. 6 (2009)
35. Zutty, J., Long, D., Adams, H., Bennett, G., Baxter, C.: Multiple objective vector-based genetic programming using human-derived primitives. In: Proceedings of the 2015 Annual Conference on Genetic and Evolutionary Computation, pp. 1127–1134. ACM (2015)

Chapter 9
Genetic Programming Based on Error Decomposition: A Big Data Approach

Amirhessam Tahmassebi and Amir H. Gandomi

Abstract An investigation of the deviations of error and correlation for different stages of the multi-stage genetic programming (MSGP) algorithm in multivariate nonlinear problems is presented. The MSGP algorithm consists of two main stages: (1) incorporating the individual effect of the predictor variables, (2) incorporating the interactions among the predictor variables. The MSGP algorithm formulates these two terms in an efficient procedure to optimize the error among the predicted and the actual values. In addition to this, the proposed pipeline of the MSGP algorithm is implemented with a combination of parallel processing algorithms to run multiple jobs at the same time. To demonstrate the capabilities of the MSGP, its performance is compared with standard GP in modeling a regression problem. The results illustrate that the MSGP algorithm outperforms standard GP in terms of accuracy, efficiency, and computational cost.

9.1 Introduction

We are entering the era of big data that refers to the explosion of available information with new promising levels of scientific exploration. Despite the novel opportunities that big data offers to recent society, it brings challenges including computational cost, huge high-dimensional sample size, storage impasse, and error extent. The rise of big data in various scientific fields such as genomics, economics, finance, neuroscience, internet security, digital humanities, etc and their challenges

A. Tahmassebi
Department of Scientific Computing, Florida State University, Tallahassee, FL, USA
e-mail: atahmassebi@fsu.edu

A. H. Gandomi (✉)
School of Business, Stevens Institute of Technology, Hoboken, NJ, USA

BEACON Center for the Study of Evolution in Action, Michigan State University, East Lansing, MI, USA
e-mail: a.h.gandomi@stevens.edu

© Springer International Publishing AG, part of Springer Nature 2018
W. Banzhaf et al. (eds.), *Genetic Programming Theory and Practice XV*,
Genetic and Evolutionary Computation, https://doi.org/10.1007/978-3-319-90512-9_9

demand new evolutionary computational paradigms to deal with salient features of big data, including heterogeneity, noise accumulation, spurious correlation, and incidental endogeneity [3]. Evolutionary algorithms have mostly been successful in solving big data problems [9, 16, 21].

Genetic programming (GP) [12] is as an extension of genetic algorithms (GA) which uses computer programs to solve problems. GP uses tree structures to represent the solutions and evolves them during generations. Prediction of real values based on each tree is the procedure by which GP performs regression [4–6, 18]. In 1998, Ryan et al. [15] proposed a relatively novel evolutionary computation method known as grammatical evolution (GE). GE provides a solution by restricting the search space using domain knowledge according to a user-specified grammar for evolving solutions. Due to the modular approach of GE, it has been successfully applied to financial applications such as predicting corporate bankruptcy, forecasting stock indices, and bond credit ratings. In addition to this, Ferreira has previously proposed another promising variant of GP, known as gene expression programming (GEP) to model nonlinear problems.

Besides the traditional tree-based GP, a linear variant of GP, know as LGP was published in Brameier and Banzhaf in 2007 [1]. The standard GP model expresses the functional programming language using tree structures in which inner nodes hold functions and leaves are the location of input predictor values. On the other hand, an evolutionary GP variant of a sequence of instructions from an imperative programming language is the essential basis for LGP. The term "linear" refers to the imperative program representation which does not mean that the method provides linear solutions [1]. Furthermore, GP has a phenomenal ability in model selection from a pool of a given population. Many of the GP-based models incorporate all the predictor input values in the modeling phase. Gandomi and Alavi [4] have previously proposed a novel scheme to formulate a problem using individual predictor variables and the interactions among them.

Different genetic operators play an essential role in the evolution process. Iba et al. [11] presented a novel method for GP, known as structured representation on genetic algorithms for nonlinear function fitting (STROGANOFF) by recombining standard GP with local hill-climbing. They have pointed out the critical changes in the semantics due to mutation operators. To overcome these difficulties: (1) they have tuned local parameters with the help of statistical identification techniques, and (2) they have controlled tree growth in GP by setting the fitness score to a minimum description length (MDL) measure. The authors validated their proposed method by comparing STROGANOFF's effectiveness in its application to symbolic regression of nonlinear problems with numerical results. Moreover, the complexity measure can be improved by tweaking the fitness function through evolution. For example, Zhang et al. [22] analyzed fitness functions on error landscapes and the complexity measures by benchmarking the importance of tree representations of GP models via a Bayesian framework. This flexibility helped to investigate the solutions to programs which might end with bloat phenomenon. In addition to this, Zhang et al. [22] improved the fitness score by balancing the complexity of the model using an adaptive learning strategy. In this procedure, the parsimony coefficient was increased to reach better accuracy. The effectiveness of their method has been tested on real-world medical diagnosis problems.

In addition to the reasonable performance of GP models in regression problems, they have also shown great performance in classification problems in various real-world and big data problems, including those in neuroscience and medical imaging. Tahmassebi et al. [19] have employed several data reduction algorithms to reduce the dimensionality of an fMRI big data classification problem. In particular, the problem with high numbers of dimensions ($\sim$240,000) was decomposed into a new problem with feasible numbers of dimensions ($<$30) via data reduction algorithms. Then, the decomposed data were used as input predictor variables for the GP classifier. Tahmassebi et al. [20] have also shown the performance of GP in classification for large numbers of generations ($\sim$13,000) using high-performance computing (HPC). This would suggest employing parallel algorithms for such population-based evolutionary algorithms to overcome the curse of dimensionality.

In this study, we propose a GP-based scheme to decompose the error in a multivariate nonlinear problem. We apply the MSGP method, previously proposed by Gandomi and Alavi [4], in solving problems with N inputs in which N 1-dimensional programs were used instead of solving the problem with one N-dimensional program. In particular, the MSGP method incorporates the individual effect of each of the input predictor variables, and the interactions among them. Additionally, it is presented that the interactions among the input predictor variables can be neglected. This decreased the computational cost dramatically without losing more than a negligible amount of accuracy. The performance of the MSGP method is tested in a problem where the deviations of the error and correlation in each stage of the MSGP method are investigated. This opens new approaches with less computational cost and the same accuracy to tackle big data problems.

9.2 Computational Model

A multi stage evolutionary algorithm, called MSGP, is presented to decompose the error through several steps. The MSGP algorithm is implemented in Python along with GPlearn and Scikit-Learn [13] libraries. The model starts with generating a population of tree-like programs to represent the data based on stochastic formulations of variables. Just a subset of the generated programs compete with each other based on the tournament size, and the winners are optimized recursively through the evolutionary process based on the fitness metrics. Three different options were set as fitness metrics: mean absolute error (MAE), mean squared error (MSE), and root mean squared error (RMSE). Additionally, the code has the ability of customizing the fitness metric by user-defined functions.

To find the best mathematical formulation and the fittest individual, different genetic operators such as crossover, subtree mutation, hoist mutation, and point mutation were employed in the GP model. To see the convergence during the evolutionary process, the size of the programs was increased which is normally expected

Table 9.1 Parameters setting for the GP and MSGP algorithms

Parameter	Setting
Population size	300
Number of generations	100
Tournament size	20
Crossover probability	0.7
Subtree mutation probability	0.1
Hoist mutation probability	0.05
Point mutation probability	0.1
Point replace probability	0.05
Parsimony coefficient	0.001
Stopping criteria	0.0
Max samples	0.9
Random state	1367
Number of jobs	1
Loss metric	MAE
Score metric	R^2
Function set	$+, -, \times, /$

to increase fitness values. Sometimes, this would never happen since it would cause computational costs and would make the final programs less understandable, a phenomenon called "bloat". To control bloat, a parsimony coefficient was defined for the GP model which made programs with lower values for the fitness metric unavailable for the selection at each generation. The other alternative was generating an offspring by applying the hoist mutation operator to insert a random subtree into the original subtree location in the next generation [14]. The MSGP algorithm was implemented in Python employing parallel algorithms to run multiple jobs at the same time. By changing the number of jobs in the code, we could use the maximum CPU cores available to decrease the computational runtime which would help to solve a big data problem. The coefficient of determination (R^2) was defined as the output score for regression problems. Table 9.1 lists the parameter setting used in the MSGP model. Figure 9.1 also presents a schematic tree structure of a program. Most of the GP models discussed in Sect. 9.1 employed all the predictor variables as inputs. This incorporation of all the variables might affect the decomposition cost throughout the modeling process. To address these issues, an MSGP strategy was proposed to model the predictor variables by taking into account the effect of each of the individual predictor variables.

The MSGP algorithm could be divided into two phases:

1. Incorporating the individual effect of the predictor variables ($MSGP_{wo-int}$).
2. Incorporating the interactions among the predictor variables ($MSGP_{w-int}$).

Fig. 9.1 A schematic tree representation of a GP model for $(\sqrt{X_1 + \frac{5}{X_2}})$

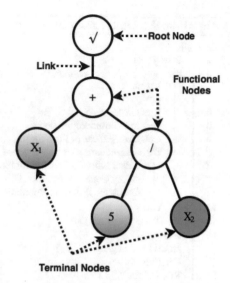

Terminal Nodes

To shine some light on the details of the MSGP algorithm, it should be noted that the final solution, $f(X)$, entails the following terms:

$$f(X) = f_1(x_1) + f_2(x_2) + \cdots + f_n(x_n) + f_{int}(X) = \sum_{i=1}^{n} f_i(x_i) + f_{int}(X) \quad (9.1)$$

where x_i is the input variable, n is the number of input variables, $f_i(x_i)$ indicates the function based on only one input variable x_i, and the interaction among the input variables was defined by $f_{int}(X)$. It is always possible to formulate a set of variables in terms of output values and a subset of the variables. Equation (9.2) presents the formulation of a binary problem with two variables based on the values predicted by the first input variable and the target output values:

$$f_2(x_2) = f(X) - f_1(x_1) \quad (9.2)$$

In other words, Eq. (9.2) demonstrates that a new variable formulates the error between the predicted and the actual values. This formulation is known as the decomposition of errors. This procedure can be extended by repeating the formulation presented in Eq. (9.2).

$$f_3(x_3) = f(X) - f_1(x_1) - f_2(x_2) \quad (9.3)$$

$$\vdots$$

$$f_n(x_n) = f(X) - f_1(x_1) - f_2(x_2) - \cdots - f_{n-1}(x_{n-1}) = f(X) - \sum_{i=1}^{n-1} f_i(x_1) \quad (9.4)$$

Algorithm 9.1 Multi-stage genetic programming (MSGP)

---

1 **begin**
2      $Y = f(X)$;
3   **for** $i = 1 : n$ *(n is the number of input variables)* **do**
        Input : x_i
        Output: Y
4   Run GP for $f_i(X_i)$;
5   Generate Initial Population ;
6   Calculate Fitness of Population ;
7   **if** *The Termination or Convergence Conditions are not satisfied* **then**
8   Select Individuals based on Fitness;
9   Apply Genetic Operators: (Crossover, Mutation,...) ;
10   Check the Fitness of Population ;
11   **end**
12         $Y \leftarrow Y - f_i(x_i)$;
13   **end**
    Input : $X (x_1, x_2, \ldots, x_n)$
    Output: Y
14   Run GP for $f_{int}(X)$;
15   Generate Initial Population ;
16   Calculate Fitness of Population ;
17   **if** *The Termination or Convergence Conditions are not satisfied* **then**
18   Select Individuals based on Fitness;
19   Apply Genetic Operators:(Crossover, Mutation,...) ;
20   Check the Fitness of Population ;
21   **end**
22     $f_{MSGP}(X) \leftarrow \sum_{j=1}^{n} f_j(x_j) + f_{int}(X)$;
23 **end**

Considering $f_{int}(X)$ presented in Eq. (9.1), the final solution calculated by the MSGP algorithm can be presented as follows:

$$f_{MSGP}(X) = \sum_{i=1}^{n} f_i(x_1) + f_{int}(X) \tag{9.5}$$

The pseudo code of the MSGP algorithm is presented in Algorithm 9.1.

9.3 Case Study

The database presented by Garzon-Roca et al. [10] was employed to compare decomposition of errors using GP and MSGP methods. The database contains experimental studies of compressive strength of masonry made of clay bricks and cement mortars. The database consists of binary inputs: (1) mortar compressive strength f_m, and (2) brick compressive strength f_b with output $f(X) = f(x_1, x_2) = f(f_m, f_b)$. Both the GP and the MSGP algorithms ran in Python [2, 13] for 100 generations and a population size of 300 to build regression models for the above-

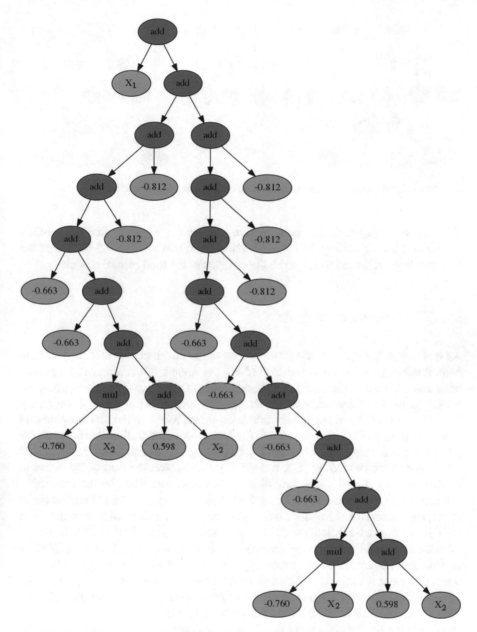

Fig. 9.2 The tree structure of the predicted solution using the GP model

mentioned database. Table 9.1 presents the details of the final parameter settings which were selected on the basis of a trial and error approach and multiple runs for the GP and MSGP algorithms. The tree structure of the resulting solution for GP is presented in Fig. 9.2. It visually presents the relation between mortar compressive

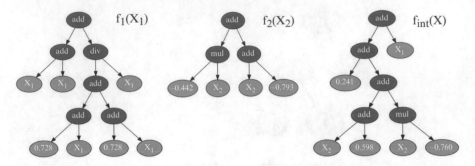

Fig. 9.3 The tree structures of the predicted solution using the MSGP model

strength ($x_1 = f_m$) and brick compressive strength ($x_2 = f_b$) to find the solution ($f(X) = f(x_1, x_2)$) using the final GP model. Additionally, Fig. 9.3 illustrates the tree structures of the solutions in the three stages of the final MSGP model.

9.4 Performance Analysis

As it is shown in Fig. 9.3, the decomposition of the errors in different stages of the MSGP method can be discussed to find out the impact of each stage in terms of error and their correlation with the outputs during the process. In this regards, the MSGP method in three different stages was considered: (1) in the first stage only the error between the actual and the predicted values based on the first variable (x_1) using GP program (f_1) was considered. (2) In the second stage, the error between the predicted values using the second input variable (x_2) and the actual values subtracted from the error calculated by (x_1), and (3) as the last stage, the sum of the errors in the first and the second stages in addition to the error calculated by the interaction function (f_{int}) were considered. In the first 10 generations, f_1 was calculated based on x_1, then from 10 to 15 generations f_2 was calculated, and from generation 15 to 25 f_{int} was used to calculate the error and R^2 score. Figure 9.4 illustrates the stages of calculating MAE through the generations. It is obvious that $f_{int}(X)$ in MSPG-II stage has just increased the accuracy of the prediction of the output values for a small amount which is infinitesimal with respect to its change in computational cost. Additionally, Fig. 9.5 also presents the R^2 score during the generations for the both GP and MSGP in the different stages. This also proves the low impact of the $f_{int}(X)$ in the MSGP$_{w-int}$ stage based on the statistical scores. More details of the calculated statistical parameters were presented in Table 9.2.

It was previously shown that a combination of minimum errors (MAE or RMSE) and $R^2 > 0.8$ lead us to a reasonable correlation between the target values and the predicted values of the models [7, 17]. To track how close the data was to the fitted regression hyperplane, R^2, the coefficient of determination for both GP and MSGP models was calculated.

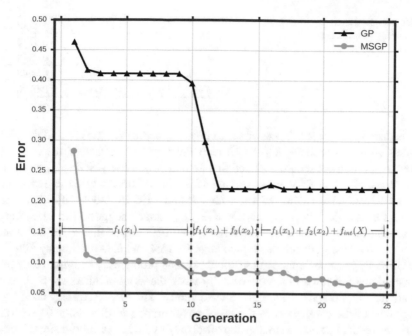

Fig. 9.4 The decomposition of the mean absolute error through the generations for the GP and the MSGP models

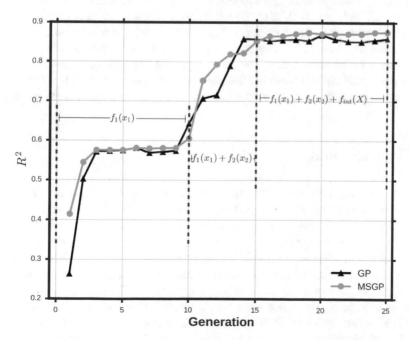

Fig. 9.5 The evolution of the coefficient of determination through the generations for the GP and the MSGP models

Table 9.2 Statistical
parameters for different
models

	GP	$MSGP_{wo-int}$	$MSGP_{w-int}$
R^2	0.8591	0.8581	0.8754
PI	0.1448	0.0469	0.0364
MAE	0.2215	0.0868	0.0648
$RMSE$	0.2791	0.0904	0.0706

As discussed, Table 9.2 presents the statistical scores for the GP and the MSGP models with and without $f_{int}(X)$. It also illustrates the effect of the interaction between variables. It should be noted that adding f_{int} in the $MSGP_{w-int}$ resulted in an increase of 2% in R^2 with respect to MSGP. In addition to this, the performance index (PI) was also calculated for both GP and MSGP models [8]. Considering the PI for both $MSGP_{wo-int}$ and $MSGP_{w-int}$ stages suggests the idea that the interaction function can be neglected in the multi-stage model. The MSGP strategy decreases the cost of decomposing the error by 15% and also runtime by 30%. The MSGP strategy can be employed to solve big data problems. The importance of the proposed method can be shown especially when the input numbers are high, where traditional GP method might not be a wise choice. The MSGP strategy changes the magnitude of the complexity of the problems from one N-dimensional problem to N 1-dimensional ones. It increases the efficiency by losing an infinitesimal increase in the accuracy and the correlation between the inputs and outputs.

Figure 9.6 depicts the 3-dimensional hyperplane solutions of the inputs by GP, $MSGP_{wo-int}$, and $MSGP_{w-int}$ models. It seems the interaction between variables changed the shrinkage path of the hyperplanes. As shown, standard GP always finds the best linear plane for the inputs data. Dealing with a binary problem brings the chance to illustrate both variables x_1 and x_2 and also the resulting regressed hyperplane in the same space. The most important aspect here would be how this hyperplane, without incorporating the interaction between the variables, could fit the best regression with a reasonable accuracy with respect to the actual output values.

9.5   Conclusions

This chapter discusses statistical and computational aspects of an efficient strategy called the MSGP method. It specifically focused on the decomposition of error and correlation in a multivariate nonlinear problem to reveal the capabilities of the MSGP algorithm to be applied to big data problems. The proposed method separates one N-dimensional problem into N 1-dimensional problems. To see how the MSGP performs, the performance of the MSGP model was compared with a standard GP model in the case of a nonlinear regression problem. The decomposition cost of both models during the generations was presented. Based on a high correlation in predicted values, the MSGP outperforms standard GP. These results suggest that the MSGP algorithm can be employed in various big data problems which are

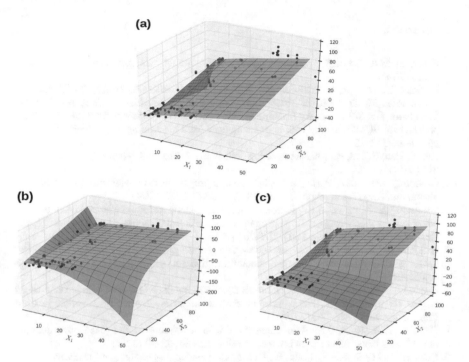

Fig. 9.6 The predicted regression hyperplanes by (**a**) GP, (**b**) MSGP$_{wo-int}$, and (**c**) MSGP$_{w-int}$ models

more difficult to solve using traditional methods due to the high computational cost. Error decreased by 15% by using the MSGP model. Additionally, the MSGP strategy reduced computational runtime by 30%. The evolution of the MAE was presented for both MSGP$_{wo-int}$ and MSGP$_{w-int}$ stages. This suggests the idea of solving the N 1-dimensional problems without considering the interactions among predictor variables. The calculated statistical scores such as R^2, and PI suggest that neglecting the interactions between the input variables causes a loss of only 1% of the correlation between the inputs and the output. This would save a reasonable amount of computational time in big data problems. To recapitulate, the MSGP method opens an innovative avenue to apply evolutionary algorithms in big data problems to overcome difficulties resulting from the big data features such as heterogeneity, noise accumulation, spurious correlation, and incidental endogeneity.

Acknowledgements This material is based in part upon work supported by the National Science Foundation under Cooperative Agreement No. DBI-0939454. Any opinions, findings, and conclusions or recommendations expressed in this material are those of the authors and do not necessarily reflect the views of the National Science Foundation. For valuable help in the revision of the chapter we also would like to thank Eitan Lees for critical and helpful comments on final draft.

References

1. Brameier, M.F., Banzhaf, W.: Linear genetic programming. Springer Science & Business Media (2007)
2. Buitinck, L., Louppe, G., Blondel, M., Pedregosa, F., Mueller, A., Grisel, O., Niculae, V., Prettenhofer, P., Gramfort, A., Grobler, J., Layton, R., VanderPlas, J., Joly, A., Holt, B., Varoquaux, G.: API design for machine learning software: experiences from the scikit-learn project. In: ECML PKDD Workshop: Languages for Data Mining and Machine Learning, pp. 108–122 (2013)
3. Fan, J., Han, F., Liu, H.: Challenges of big data analysis. National Science Review 1(2), 293–314 (2014)
4. Gandomi, A.H., Alavi, A.H.: Multi-stage genetic programming: a new strategy to nonlinear system modeling. Information Sciences 181(23), 5227–5239 (2011)
5. Gandomi, A.H., Alavi, A.H.: A new multi-gene genetic programming approach to nonlinear system modeling. part II: geotechnical and earthquake engineering problems. Neural Computing and Applications 21(1), 189–201 (2012)
6. Gandomi, A.H., Alavi, A.H.: A new multi-gene genetic programming approach to nonlinear system modeling. part I: materials and structural engineering problems. Neural Computing and Applications 21(1), 171–187 (2012)
7. Gandomi, A.H., Alavi, A.H., Mirzahosseini, M.R., Nejad, F.M.: Nonlinear genetic-based models for prediction of flow number of asphalt mixtures. Journal of Materials in Civil Engineering 23(3), 248–263 (2010)
8. Gandomi, A.H., Roke, D.A.: Assessment of artificial neural network and genetic programming as predictive tools. Advances in Engineering Software 88, 63–72 (2015)
9. Gandomi, A.H., Sajedi, S., Kiani, B., Huang, Q.: Genetic programming for experimental big data mining: A case study on concrete creep formulation. Automation in Construction 70, 89–97 (2016)
10. Garzón-Roca, J., Marco, C.O., Adam, J.M.: Compressive strength of masonry made of clay bricks and cement mortar: Estimation based on neural networks and fuzzy logic. Engineering Structures 48, 21–27 (2013)
11. Iba, H., deGaris, H., Sato, T.: A numerical approach to genetic programming for system identification. Evolutionary Computation 3(4), 417–452 (1995).
12. Koza, J.R.: Genetic programming: on the programming of computers by means of natural selection, vol. 1. MIT Press (1992)
13. Pedregosa, F., Varoquaux, G., Gramfort, A., Michel, V., Thirion, B., Grisel, O., Blondel, M., Prettenhofer, P., Weiss, R., Dubourg, V., Vanderplas, J., Passos, A., Cournapeau, D., Brucher, M., Perrot, M., Duchesnay, E.: Scikit-learn: Machine learning in Python. Journal of Machine Learning Research 12, 2825–2830 (2011)
14. Poli, R., Langdon, W.B., McPhee, N.F., Koza, J.R.: A field guide to genetic programming. Lulu. com (2008)
15. Ryan, C., Collins, J., Neill, M.: Grammatical evolution: Evolving programs for an arbitrary language. In: European Conference on Genetic Programming, Paris 1998, pp. 83–96 (1998) Springer, Berlin (1998)
16. Schadt, E.E., Linderman, M.D., Sorenson, J., Lee, L., Nolan, G.P.: Cloud and heterogeneous computing solutions exist today for the emerging big data problems in biology. Nature Reviews Genetics 12(3), 224–224 (2011)
17. Smith, G.N.: Probability and statistics in civil engineering. Collins Professional and Technical Books 244 (1986)
18. Tahmassebi, A., Gandomi, A.H.: Building energy consumption forecast using multi-objective genetic programming. Measurement 118, 164–171 (2018)
19. Tahmassebi, A., Gandomi, A.H., McCann, I., Schulte, M.H., Schmaal, L., Goudriaan, A.E., Meyer-Bäse, A.: An evolutionary approach for fMRI big data classification. In: 2017 IEEE Congress on Evolutionary Computation (CEC) pp. 1029–1036 (2017)

20. Tahmassebi, A., Gandomi, A.H., Meyer-Bäse, A.: High performance GP-based approach for fMRI big data classification. In: Proceedings of the Practice and Experience in Advanced Research Computing 2017 on Sustainability, Success and Impact, PEARC17, pp. 57:157:4. ACM Press, New York, NY, USA (2017)
21. Wu, X., Zhu, X., Wu, G.Q., Ding, W.: Data mining with big data. IEEE transactions on knowledge and data engineering **26**(1), 97–107 (2014)
22. Zhang, B.T., Mühlenbein, H.: Balancing accuracy and parsimony in genetic programming. Evolutionary Computation **3**(1), 17–38 (1995)

Chapter 10
One-Class Classification of Low Volume
DoS Attacks with Genetic Programming

Stjepan Picek, Erik Hemberg, Domagoj Jakobovic, and Una-May O'Reilly

Abstract We use Genetic Programming in a machine learning approach to learn a detector of DoS-related network intrusion events. We present a one class classifier technique that trains a model from one class of data—normal, i.e., non-intrusion events. Our technique, after ensemble fusion, is competitive with one-class modeling with Support Vector Machines. We compare with three datasets and our best GP-based classifiers are able to outperform one-class SVM. For two out of four test cases, the advantage of GP classifiers when compared with one-class SVM is less than 1% which does not represent a significant improvement. On the last two cases, GP achieves significantly better results and making it a viable choice for anomaly detection task.

10.1 Introduction

Denial of Service (DoS) cyber attacks present a serious threat to computer systems and inflict significant economic damage. They disrupt critical public and enterprise services. DoS attacks can be characterized by their *attack surface*, e.g. application resources, protocol or network, by their *volume*, and in terms of *how they are measured*: bandwidth magnitude is measured in bits per second (Bps), protocol layer attacks are measured in packets per second and application layer attacks are measured in requests per second. Many DoS attacks are advanced by malicious network intrusion events that flood a system's resources so that services to legitimate requests are denied.

S. Picek (✉) · E. Hemberg · U.-M. O'Reilly
MIT, CSAIL, Cambridge, MA, USA
e-mail: stjepan@computer.org; unamay@csail.mit.edu

D. Jakobovic
University of Zagreb, Faculty of Electrical Engineering and Computing, Zagreb, Croatia
e-mail: domagoj.jakobovic@fer.hr

© Springer International Publishing AG, part of Springer Nature 2018
W. Banzhaf et al. (eds.), *Genetic Programming Theory and Practice XV*,
Genetic and Evolutionary Computation,
https://doi.org/10.1007/978-3-319-90512-9_10

149

One such example is the so-called SYN Flood attack [6]. It exploits a known vulnerability in the TCP (Transfer Connection Protocol) connection sequence. This sequence has three steps. (1) Host A sends a SYN request to open a connection to host B. (2) B then responds with a SYN-ACK response and waits while holding resources for A. (3) A confirms with an ACK. In a SYN Flood attack, the requester A sends multiple SYN requests but either does not respond to B's SYN-ACK response, or sends the SYN requests from an IP address that is not its own (*spoofed*). Either way, B continues to wait for each ACK response so eventually no new connections can be made and ultimately denying B from providing connection services. Other DoS flood attacks take advantage of similar ways to tie up a resource. Low volume DoS flood attacks, including SYN Flood, rely on flying under the radar, i.e., sparsely displaying a signature in network traffic flows, in order to evade intrusion detection techniques [12]. In this paper we focus on intrusions closely related to DoS attacks. Detecting such attacks is often very difficult but highly valuable because DoS attacks can be high volume.

One option stemming from the nature-inspired computation area for developing intrusion classifiers is to use Genetic Programming (GP). GP has been used to learn a binary classifier that discriminates between the normal and the anomalous data a number of times with good results [7]. When the problem offers only one class (i.e., normal data) some researchers have approached the problem by synthetically creating a second, "non-normal", class of data and continuing to use GP to learn a binary classifier [5]. Others have approached this problem from the unsupervised learning perspective [13].

Differing from this approach, herein we present the design of a GP intrusion classifier (i.e., a detector) that requires only one class to regress. We train the classifier with only normal data by selecting for mappings with outputs that lie within a restricted range. When we evaluate candidate classifiers, in the testing phase, we test with both normal and anomalous data. Afterward we ensemble selected classifiers from our runs and perform further evaluations. In order to evaluate its efficiency we compare our method with one-class SVM on several datasets (network logs) that are either created for the purpose of benchmarking a classifier or taken from practical settings. Our one-class classification technique is general, i.e., it extends to problems of anomaly outside intrusion detection.

Our main contributions are:

- Design of a new one-class GP classifier using symbolic regression and interval mapping.
- Design and evaluation of ensemble classifiers based on the one-class GP classifier.
- Experimental evaluation of our approach on a number of datasets and comparison with one-class SVM.

The rest of this paper is organized as follows. In Sect. 10.2 we present the design of our GP-based one-class classifier and outline our ensemble fusion method. In Sect. 10.3 we summarize related work. Section 10.4 presents the datasets we use, experimental settings, results, and discussion. Finally, in Sect. 10.5 we consider possible future research and conclude.

10.2 Our Method

An adversarial model defines the scope of intrusions a classifier is expected to detect. Our model assumes the adversary is able to launch attacks identifiable as sourced from any IP address and can, prior to intrusion, conduct a reconnaissance phase to identify its target. The adversary can launch different types of attacks, not only DoS which imply it seeks to penetrate the network perimeter of the target. Intrusions outside this model are not considered. One such example is an adversary who can influence the GP training process. We further assume that we only need to collect data from a network that has not experienced intrusion providing us with data of class "normal".

10.2.1 Intuition

We ask GP to evolve a mapping that compresses normal class data into a pre-defined target interval $I = [lower \ldots upper]$. An output for an input datum that is anomalous should be mapped outside I. On an intuitive level, we seek a compressive relation that embeds data from a high dimensional feature space to a one-dimensional space with a specific "normal" interval for normal class data and values outside the interval for "anomalous" data. We want the interval to be as small as possible to minimize both false negatives, i.e., anomalous data mapping into the interval, and false positives, i.e., normal data falling outside it.

Since GP computes floating point numbers, we use how floating points are stored in computers and their discretely decreasing density to define the target intervals for a classifier. For floating point representation, computers use a system highly resembling that of exponential notation where a number can be stored in binary notation (recall any number can be expanded to the binary representation) as $\pm n \times 2^E$ with n being the significand, E the exponent, and $1 \leq n < 2$. A 32-bit word can then be divided into 1 bit for the sign, 8 bits for the exponent, and 23 bits for the significand. Since the significand has 23 bits, the gap (precision) between the floating point value 1 and the next larger floating point (in binary, first bit equal to 1, followed by 21 zero bits and the last bit equal to one) is $\eta = 2^{-22}$. That gap between adjacent floating point numbers becomes bigger as the magnitudes of the numbers become bigger, and smaller as the magnitudes of the numbers become smaller.

We illustrate the density of floating point numbers in a small example in Fig. 10.1. The precision as well as the magnitudes of the floating point numbers are defined by the IEEE floating point representation [1]. Additional details about the floating point and how are they stored in computers can be found in [15].

Fig. 10.1 Density of floating point values on the real line

-4   -2   -1   0   1   2   4

Table 10.1 Classification outcomes for $\hat{y} = GPF^R(X_i)$

Outcome	Description
TP	$\hat{y} = GPF(X_i; R) \notin R$ and class$(X_i) = 1$
FN	$\hat{y} = GPF(X_i; R) \in R$ and class$(X_i) = 1$
TN	$\hat{y} = GPF(X_i; R) \in R$ and class$(X_i) = 0$
FP	$\hat{y} = GPF(X_i; R) \notin R$ and class$(X_i) = 0$

In our approach, we treat the choice of the target interval as a parameter of the learning method and we use intervals [0, 1], [1, 2], [2, 3], [3, 4], [4, 5], [7, 8], [8, 9], [15, 16], [16, 17], [31, 32], and [32, 33]. These intervals are chosen because either the lower or the upper boundary coincides with an increasing power of 2, since the powers of 2 mark the change in the density of floating point representation. We ask GP to evolve classifiers for smaller and smaller intervals. We expect GP to find larger intervals more easily, e.g. forcing the output for all training instances to [0, 1] could be done trivially, but the interval [16, 17] requires a mapping with a smaller range. After a sweep through different target intervals, we select a classifier from the smallest interval where we obtain acceptable training performance.

We use absolute values so we consider both signs (negative and positive); range $[0, 1 >$ (or $< -1, 1 >$ if not taking absolute value) holds half of all IEEE floating point numbers, which is approximately 2^{63}, while the range $[1, inf >$ holds the other half. Range [1, 2] holds approximately 2^{53} numbers, which are all the combinations using the same exponent. Ranges [2, 3] and [3, 4] both hold 2^{52} numbers, [4, 5] and [7, 8] hold 2^{51} etc.

10.2.2 Formal Definition

Our GP evolves a function GPF in the form of a regression tree that produces a single value. Let GPF be a function (classifier) evolved by GP and R the set of intervals we regress our values to, i.e., $R \in [0, 1], [1, 2], [2, 3], [3, 4], [4, 5], [7, 8], [8, 9], [15, 16], [16, 17], [31, 32], [32, 33]$. Next, n denotes the number of instances in the dataset and X is the vector of m features – $(x_1, \ldots, x_m)$. Y is the set of all possible class labels, here $Y = (0, 1)$, $y \in Y$ is the actual label of instance Xi, and $\hat{y}$ is the predicted label of X.

Our goal is to learn the classifier $y = f(X_i; R)$ where R denotes the target interval. The predicted label for $\hat{y}$ is "normal" for $\hat{y} = GPFx(X_i; R) \in R$ and "anomaly" for $\hat{y} = GPF(X_i; R) \notin R$. In Table 10.1 we present how to derive classification outcomes. The normal class is denoted as negative ("0") and anomaly as positive ("1"), but the choice of labels is arbitrary.

Note that in the training phase, which includes only normal instances, only true negatives and false positives are possible and we evaluate a model's accuracy as follows:

$$ACC_1(GPF^I) = \frac{TN}{TN + FP}.$$ (10.1)

There are a number of design options to supplement the method's fitness function to express more than accuracy. One option is to posit that all the features of the data are relevant to the unseen "anomalous" class and therefore GP should use all (or at least the majority) of input features. Fitness pressure to express this is achieved by multiplying the classifier accuracy with the *fraction of the features* from the entire input set that appear in the GP model. This should further serve to prevent GP from evolving "cheat" solutions that are trivial mappings that don't depend on the input at all.

The resulting training fitness function for the GP is:

$$fitness = ACC_1 \cdot \frac{\texttt{nUsedFeatures}}{\texttt{nAllFeatures}},$$ (10.2)

Here, `nUsedFeatures` is the number of features that appear in the tree and `nAllFeatures` is the total number of features in the dataset. With this fitness function, only a solution including all the features can have the best possible fitness. Another design option could maximize the range of the mapped outputs for the training set or the number of unique outputs. Our experiments use the first design option. We check whether some features in the evolved classifier are used in a trivial way, e.g., a feature that is subtracted from itself, and we do not include such features in the `nUsedFeatures` number. Naturally, it is still possible that although all features are used in the tree that some of them are actually canceled out.

After each generation of GP with the training set, we check the best classifier on cross-validation split and stop immediately when the cross-validation score gets worse. We report the accuracy of an evolved classifier using test data distinct from the training or cross-validation splits, composed of both normal and anomaly instances. Our measure is:

$$ACC_2(GPF^I) = \frac{TP + TN}{TP + TN + FN + FP}.$$ (10.3)

We also report the $F1$ measure on the test data:

$$F1 = 2\frac{precision \cdot recall}{precision + recall},$$ (10.4)

where *precision* is the number of correct positive results divided by the number of all positive results, while recall is the number of correct positive results divided by the number of positive results that should have been returned.

We do not have a feature selection step within our method. While fewer features will make classification faster and may help with training set accuracy we anticipate some features that are not relevant to discriminating "normal" may be relevant for "anomaly". Feature selection only makes sense when learning a binary classifier from this perspective. We present the pseudocode for our GP classifier in Algorithm 10.1.

Algorithm 10.1 One-class GP classifier
___
Input:
R – set of ranges,
S_{train} – training set,
S_{xval} – cross-validation set,
S_{test} – testing set,
Output:
GPF^r – evolved classifier,
$ACC_1(GPF^r, S_{train})$ – accuracy on the training set for the evolved classifier,
$ACC_2(GPF^r, S_{test})$ – – accuracy on the testing set for the evolved classifier,
$F1(GPF^r, S_{train})$ – F1 measure on the testing set for evolved classifier,
$fitness(GPF^r, S_{train})$ – fitness score for the training set and evolved classifier,
repeat
   $r = next\ in\ R$
   repeat
      $find\ best\ classifier\ that\ maximizes\ fitness\ on\ training\ set$
     until $ACC_1(GPF^r_{current}, S_{xval}) < ACC_1(GPF^r_{previous}, S_{xval})$
until $all\ ranges\ tested$

10.2.3 Ensemble Formation

Since GP produces a population of solutions, and since we execute the GP in several runs, a natural question is whether it is possible to use more than one solution in order to obtain more reliable results. Consequently, we use GP as an ensemble classifier where the size of ensemble varies. We note that GP ensembles were also used before for intrusion detection frameworks but the GP part was understandably different from ours [8].

To construct an ensemble, we simply choose some among the evolved models from multiple runs. The models are included in the ensemble in the order of their decreasing output standard deviation, up to the target ensemble size. To obtain the fused result, we either use voting (where we must use odd number of models) or compute the average output value before determining the instance class. To conclude, in order to use ensemble formation, we need to select the following parameters:

1. The target range selection policy.
2. The model selection policy and the number of classifiers in the ensemble.
3. The prediction fusing policy.

10.3   Related Work

Anomaly based detection is a well researched topic in the last decade and more with many papers examining various defense types or algorithms to be used. In this section, we give only a short overview of relevant works in order to better understand the variety of approaches used up to now. Intrusion detection techniques are usually divided into **signature based** and **anomaly detection based** approaches. In the signature based approaches one relies on recognizing the signatures of attacks (e.g. hash values that are characteristic for certain attack types). Such detection techniques are easily avoided by modifying the attack or using previously unknown attack (zero-day attack). Anomaly detection systems rely on recognizing what is normal traffic and categorizing all that does not fit the description of normal into anomaly.

One-class GP is an idea introduced by Curry and Heywood where they artificially create the second class (outliers) on the basis of the normal data that is available [5]. We note that we do not consider it to be appropriate for network anomaly detection scenarios since one can only create data that does not belong to the normal data that are available, which does not mean that such created data correspond to anomalies.

Cao et al. experiment with one-class classification by using kernel density function where the density function is approximated by using genetic programming symbolic regression [3]. Their results improve over standard one-class KDE and the authors report good results on a number of datasets where one is the KDD Cup dataset. We construct our one-class GP differently, where we do not artificially create anomaly data. In addition, we do not use feature selection when training the one class GP.

To and Elati developed a one-class GP where they use only one class in the training [22]. In their approach, GP tries to find a curve that fits all patterns in the training set. Next, patterns close to the curve are selected where the proximity is evaluated with Euclidean distance. Then, if an instance belonging to the testing set is close to the trained patterns, it is defined as belonging to the normal class.

Orfila et al. use genetic programming in order to train easy-to-understand network intrusion detection rules [14]. The authors concluded that GP can be used to generate short rules that are easily understood which facilitates the understanding of the semantics of the attack.

Song et al. use genetic programming to detect anomalies in KDD Cup dataset where the authors use the hierarchical dynamic subset selection in order to be able to train on around 500,000 instances [20].

There has been also a series of work concentrating on the feature selection for the anomaly detection, see e.g. [24, 26]. Still we note that it is hard to conduct feature selection with anomaly detection since the important features for the normal class do not necessarily need to be important features for anomaly class.

For a somewhat outdated but extensive overview of computational intelligence methods for usages in the intrusion detection system we refer interested readers to [25]. For an overview of machine learning techniques for intrusion detection we refer readers to [23]. Finally, for a general reference on outlier analysis, we refer readers to [2].

10.4 Experiments

In this section, we first describe the datasets we use for evaluation, our experimental settings, and the algorithm to which we compare. Then, we present results for one-class classification.

10.4.1 Datasets

We evaluate with three datasets named KDD, NSL-KDD, and Proprietary. KDD is the oldest and was created from manually generated traffic in a controlled (research) network. The traffic has been criticized as unrealistic [19] and some data is redundant. The NSL-KDD dataset is a revision of KDD to address these problems. Both datasets are in common use as benchmarks so we use them. They also offer examples of multiple attack types which is not always possible to obtain from a single real dataset. The third dataset is proprietary. It consists of only one attack type and normal traffic. This type of attack is also present in the KDD and NSL-KDD datasets.

10.4.1.1 KDD Cup Dataset

The KDD Cup dataset [21] is extracted from 9 weeks of raw TCP dump data for a local-area network (LAN) simulating a typical U.S. Air Force LAN being exposed to multiple attacks. The dump consists of about five million connection records selected for training purposes and around two million connection records for testing purposes. The records are grouped into sequences of TCP packets starting and ending at some well defined times. Each sequence can be labeled as either normal or anomalous. Each sequence has 41 features [21]. Anomalous sequences can be further divided into four classes:

1. **DoS**—denial-of-service attacks.
2. **Probe**—surveillance and other probing attacks.
3. R2L—unauthorized access from a remote machine.
4. U2R—unauthorized access to local superuser privileges.

We use 25,000 instances in the training set and 5500 in the cross-validation set, all normal. One testing set is comprised of 75,000 instances, normal and anomalous.

The anomalous class includes instances of two different attack types: DoS and Probe grouped under the "anomaly" label. In the testing set, 35% of instances belong to the anomaly class. We could also include all instances of every attack type. We call this second KDD dataset KDD^* and it consists of 77,000 instances in the testing set (training and cross-validation sets are the same as for the KDD Cup dataset). For the KDD^* testing set, 38% of instances belong to the anomaly class.

10.4.1.2   NSL-KDD Dataset

The second dataset is the NSL-KDD which attempts to remedy some of the problems of the KDD Cup dataset [9]. The differences are:

- The dataset does not include redundant records in the train set.
- There are no duplicate records in the proposed test sets.
- The number of selected records from each difficulty level group is inversely proportional to the percentage of records in the original KDD Cup dataset.
- The number of records in the train and test sets are smaller (which enables us to use all the instances).

The details for this dataset are the same as for the KDD Cup dataset (i.e., the same number of features) and we use 10,000 instances in the training, 3500 in the cross-validation, and 22,000 instances in the testing phase. Testing set has 57% of instances belonging to the anomaly class.

10.4.1.3   Proprietary Dataset

The last dataset we use for evaluation is a proprietary dataset obtained from an anonymized Internet provider. In order to preserve the confidentiality of data, we report only its basic characteristics. It consists of only 9200 network logs with 3000 instances belonging to the normal traffic in the training set and 700 instances belonging to the normal traffic in the cross-validation set. As before, instances are starting and ending at some well defined times in accordance to the rules as defined by the firewall in use. Post-hoc analysis showed the remaining instances to be a type of low intensity DoS attack called Syn Flood, see Fig. 10.2. It is possible that this dataset has instances labeled incorrectly, most likely false positives ("normal" labels for anomalous data). This cannot happen with the manually generated datasets NSL-KDD and KDD. Each record consists of 15 features extracted from the raw data. We do not conduct any feature selection since there is no a priori knowledge on what features are the most important. Proprietary testing set has 31% of instances belonging to the anomaly class.

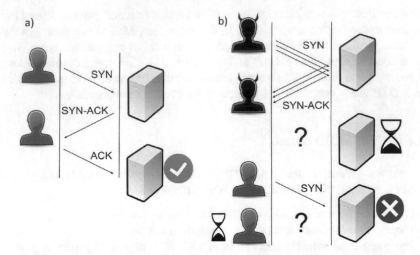

Fig. 10.2 Syn Flood attack. Part (**a**) Normal traffic between legitimate user and server. Part (**b**) The attacker sends several packets but does not send the "ACK" back to the server. The connections are half-opened and consuming resources on server. A legitimate user tries to connect but the server refuses to open a connection resulting in a denial of service

10.4.2 Genetic Programming Parameters

The input terminals are the features of the dataset and they are treated as real values. To include the nominal features with discrete values, these have been mapped to the set $\{0, 1, \ldots\}$ containing as many values as the given nominal feature. The rest of the features are used without any transformation.

As functions, we use the standard arithmetic binary operators $(+, -, *, /$ (protected)), as well as the square root function and the branch operator *iflte*, which accepts four arguments, returns the third argument if the first one is less than or equal to the second, and the fourth one otherwise. The square root and division operators are protected so that the square returns 0 if the argument is negative, and division returns 1 if the denominator is close to zero.

For all the GP classifiers, the training phase is conducted with a population size of 500 individuals; all the training combinations are executed in 30 runs (repetitions). In the evolution process, GP uses a 3-tournament selection, where the worst of the three randomly selected individuals is eliminated. A new individual is immediately created by applying crossover to the remaining two individuals from the tournament. The new individual is then mutated with a probability of 0.5. The crossover is performed with five different tree-based crossover operators selected at random: a simple tree crossover with 90% bias for functional nodes, uniform crossover, size fair, one-point, and context preserving crossover [16]. The mutation operators are subtree, shrink, hoist, permutation, and Gaussian mutation of ephemeral random constants, applied at random for each mutation operation. The GP implementation is based on the Evolutionary computation framework [10].

10.4.3   Comparison Algorithm

We compare to one-class SVM. SVM is a semi-supervised learning algorithm where the support vector model is trained on instances belonging to only one class [18]. That class is usually called "normal" class and the one-class SVM tries to infer the properties of that class and from them predict which examples are not like the normal class, i.e., they are anomalies. One-class SVM is therefore usually used for anomaly detection due to the fact that the lack of training examples is what defines anomalies. The one-class classification is reached by searching a hyperplane with a maximum margin between the target data and the origin. Note that since SVM decision boundaries are soft, it can be used as an unsupervised algorithm as well. The implementation we use is from LIBSVM [4] as available in the R tool [17]. Further details about one-class classification techniques can be found in [11].

We perform a tuning phase of the SVM parameters for each dataset. In all our experiments we use a radial basis kernel and tune ν and γ parameters. Here, ν parameter is an upper bound on the fraction of margin errors and a lower bound of the fraction of support vectors relative to the total number of training examples. The γ parameter defines how far the influence of a single training example reaches. The parameter values resulting from the tuning phase are given in Table 10.2.

10.4.4   Classification Results

In the GP case, we varied the target output interval in the learning phase, and the same interval is used in the testing phase: if an instance in the testing phase is mapped outside the given interval, it is classified as the anomaly. To assess the efficiency of our GP classifier, we consider two measures; accuracy and F1 measure. In this case, the accuracy measure should be considered only as a rough indication of classifier behavior and not as a definitive measure for assessing the performance of a classifier. This is especially correct in scenarios where anomaly data is much less represented than the normal data since then even trivial classifiers would attain a high accuracy by simply putting all measurements in the normal class.

Table 10.2 One class SVM parameter tuning

Parameter	KDD/KDD*	NSL-KDD	Proprietary
ν	0.001	0.001	0.5
γ	0.1	0.1	0.001

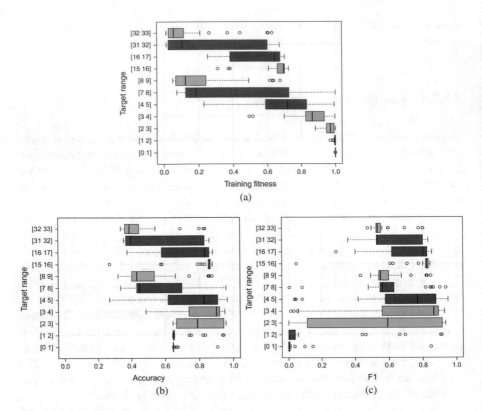

Fig. 10.3 GP one-class regression, KDD Cup dataset. (**a**) Training results. (**b**) Test results (accuracy). (**c**) Test results (F1)

The training results on the KDD Cup dataset for the GP classifier are given in Fig. 10.3a. We observe that the GP easily succeeds in producing models which include all of the features (since the maximum fitness equals to 100%) and still output values in the desired interval. This is especially true for ranges [0, 1], [1, 2], and [2, 3] where it is trivial to fit almost all instances into those ranges. We note that in cases where fitness is lower, this is almost in every instance due to the *accuracy* term in Eq. (10.2), while the second term equals 1, which means that the GP is able to include all the features easily. As the most interesting ranges we consider [3, 4], [4, 5], [15, 16], [16, 17] where the fitness value is still high and behavior is stable.

The testing results are given in Fig. 10.3b for the accuracy and Fig. 10.3c for the F1 measure. While in the training results the models are able to accurately map the training data with absolute precision for some target ranges, we can see from the test results that these ranges do not provide models with generalization capabilities. The results indicate that the most of the evolved classifiers tend to classify all the testing instances as belonging to the normal class; this result is the consequence of forcing the output to the desired range in the training phase. For [0, 1] and [1, 2] ranges,

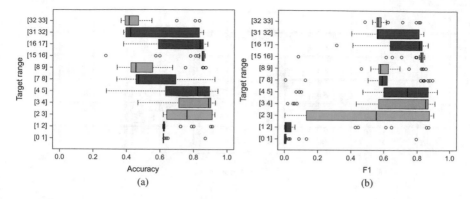

Fig. 10.4 GP one-class regression, KDD^* dataset. (**a**) Test results (accuracy). (**b**) Test results (F1)

accuracy is around 65% which actually represents the percentage of normal data in the whole dataset. Consequently, F1 measure reveals the problems with those ranges where we see the results to be 0.

Rather than using the feature information in a meaningful manner, some of the developed models simply "play it safe" and retain the target output value range regardless of the input features. The same result can be observed in the case of the one-class SVM, which also exhibits this behavior, classifying the majority of test instances in the normal class (Table 10.3). Since the evolved models do not show a great difference in training fitness values, the open question is the identification of model properties that would indicate better performance on the unseen data.

The same models that were trained using the KDD Cup dataset are tested on the KDD^* dataset, which uses the same features. The test results in terms of accuracy and F1 measures are shown in Fig. 10.4a and b, respectively. It can be seen that the similar level of accuracy is reached for the same target ranges as in the KDD Cup dataset. Likewise, ranges [0, 1] and [1, 2] result in trivial classifiers where all data is classified as normal.

Next, Fig. 10.5a–c give results for the NSL-KDD dataset for training, testing with accuracy, and testing with F1 measure, respectively. In the training phase the smallest ranges again easily fit all the data.

The final set of results are given for the proprietary dataset, with training results given in Fig. 10.6a. The testing results are shown in Fig. 10.6b for accuracy and in Fig. 10.6c for the F1 measure. We can observe that in this dataset the test performance over different target ranges is much more diverse, and it is difficult to reach a conclusion of the most effective range.

In Table 10.3 we give results for testing phases for GP as well as for one-class SVM. We note that for all investigated datasets, GP was able to reach higher accuracy and F1 measure than the one-class SVM. We also depict those results in Fig. 10.7a and b for accuracy and F1 score, respectively. Still, the GP efficiency varies considerably over multiple training runs, which is an issue we address further in the next section.

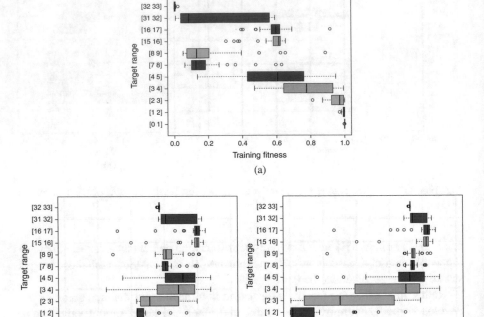

Fig. 10.5 GP one-class regression, NSL-KDD dataset. (**a**) Training results. (**b**) Test results (accuracy). (**c**) Test results (F1)

Finally, we give an example of a solution (in prefix notation) obtained with GP for one-class classification for the KDD Cup dataset.

```
iflte((var15+var27)*(var11-var18),iflte(iflte(var10,
var2,var21,var34),iflte(var1,var20,var12,var17),
sqrt(var13),iflte(var16,var3,var14,var30)),iflte(var7,
var8,var39,var41)*(var26-var9),(var28/var27)+
(var36*var6))-(var38+var37-var23/var14-var24/var22)+
sqrt(iflte(var25,var15,var37,var16)+sqrt(var33)-
iflte(iflte(var5,var19,var32,var33),var40-var35,
var4*var29,var31*var25))
```

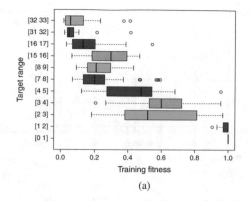

(a)

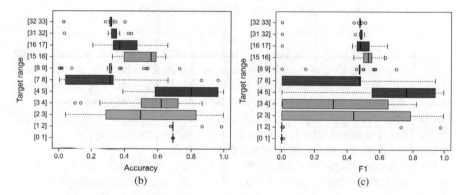

(b)   (c)

Fig. 10.6 GP one-class regression, proprietary dataset. (**a**) Training results. (**b**) Test results (accuracy). (**c**) Test results (F1)

Table 10.3 Testing results (accuracy/F1)

Algorithm		KDD	KDD*	NSL-KDD	Proprietary
One-class SVM		95.70/94.10	93.70/91.70	80.90/81.20	83.30/73.80
GP [4, 5]	min	26.94/3.62	27.73/6.82	34.73/16.31	38.87/0.00
	max	96.49/94.99	94.35/92.27	87.83/89.96	99.69/99.50
	median	82.69/76.85	82.20/74.02	71.12/72.27	79.92/75.59

10.4.5   Selection of GP Models and Ensemble Classifier

From the results in the previous section we can observe that GP is able to produce highly accurate classifiers. However, there is unfortunately no visible correlation between the training fitness of a model and its efficiency on the test dataset, since there are models with high fitness and poor generalization properties. Likewise, a lower training fitness model can still obtain significantly better result on test instances. Therefore, one needs to determine both which target range and which model to use for unseen instances.

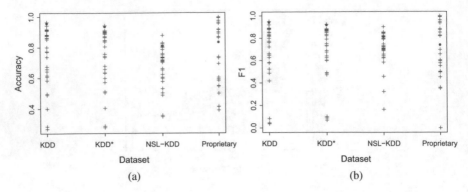

Fig. 10.7 GP (in blue plus symbol) with target range [4, 5] vs one-class SVM (in red filled circle symbol). (**a**) GP vs one-class SVM (accuracy). (**b**) GP vs one-class SVM (F1)

As for the target range, we can see from the results in the previous section that ranges below [4, 5] should be avoided because the evolved models do not have a satisfactory generalization ability. This is also evident from examining the individual models, which in lower ranges have a very small deviation of output values, or a very small number of unique outputs (e.g., all the instances are mapped to a single output value). These are trivial models which do not actually depend on input data but form an identity function to fit the desired training interval.

For the higher ranges the training fitness decreases as the distance from zero is greater, making it harder to evolve a model that would capture the training instances. Therefore, any as high range as possible that offers high enough training fitness could be considered for use.

To chose a target range relying solely on the training data, we design a selection policy where we use the range that has the largest number of highly fit individuals. In order to select it, we could for instance use the median value or the number of models above some minimal classification score σ where that parameter needs to be user-defined. For all the datasets in the previous section, based on the training results in Fig. 10.3a, b, this would result in the range [4, 5] as the chosen one; consequently, we will concentrate on this range in further analysis.

As for the choice of a particular model, simply choosing the one with the highest training fitness does not guarantee the most successful performance on the test set. On the contrary, often the models with perfect training score perform poorly on unseen instances. The same behavior is present with the accuracy on the cross-validation data, which also does not show meaningful indication of test performance. Here, the goal would be determining whether the training phase produced a "regular" classifier or just a trivial arithmetic operation to map all the input combinations into the given output interval. One way of performing this

Table 10.4 Training model selection—KDD Cup, range [4, 5]

Training fitness	0.6586	0.5587	0.2332	0.4336	0.92584	0.4069	0.66032	0.9042
F1 (test data)	83.7	87.9	3.6	88.6	93.1	88.6	84.3	76.3
St. dev. (training)	1.8×10^6	1.5×10^6	1.1×10^6	200,227	157,564	69,228	59,717	24,795

Table 10.5 Ensemble test results—all datasets, range [4, 5] (F1 measure)

Dataset	KDD		KDD*		NSL-KDD		Proprietary	
Ensemble size	Vote	Average	Vote	Average	Vote	Average	Vote	Average
3	84.5	39.3	83.2	42.8	79.5	63.4	94.3	60.1
7	87.9	59.9	87.4	62.1	85.3	75.1	90.4	57.1
11	91.3	60.2	89.6	62.1	84.4	81.5	65.0	55.3
15	89.7	54.1	88.7	56.2	86.9	78.5	80.0	56.4

analysis is to test the variability of output values, thus forcing the model to include the information in a meaningful way. To achieve this, we analyze the models' behavior on the training set, recording the GP output values for all training instances, and calculate the *standard deviation* of those values. We note that we do not presume any specific distribution of output values, but merely aim to estimate the model efficiency.

We therefore concentrate on all evolved models and sort them not by training fitness, but in the *decreasing* order of standard deviation of the values they produce. A sample of these models for the KDD Cup dataset in the range [4, 5] is shown in Table 10.4. While not perfect, the selection of models based on this criterion (greater values of deviation) provides a more reliable outcome than relying on training fitness only. This simple analysis also immediately reveals a number of models with perfect training score but with deviation close to zero (or exactly zero) which can be quickly discarded.

If we follow the guidelines for the range, we construct the ensembles using the range [4, 5]; the final results for all datasets are given in Table 10.5. Regarding the ensemble construction methods, we can immediately note that the averaging method should not be considered; this is the consequence of different models using radically different output values which cannot be added in a meaningful manner.

The obtained results are encouraging since we are able to increase the performance of our GP classifier, where we note that the best results should be obtained with medium sized ensembles (of 7 or 11). At the same time, there is almost no overhead stemming from using GP in ensemble setting. Indeed, since GP works with populations we already have the solutions and the only additional step is sorting them and running the voting or averaging. However, for effective ensembles of target size, the total number of runs should be significantly higher than the ensemble size to provide diversity of models to choose from.

10.5 Conclusion

In this paper, we investigate how to employ genetic programming as one-class classifier in order to detect anomalies in the network traffic. To be able to use GP as one-class classifier we use it in a regression style where all anomaly instances fall out of a specified range. To vary the difficulty of regressing, we use the fact that density of floating point numbers to real values decreases the farther away from zero one goes. The results indicate that the GP is able to cope with the investigated problems and exhibits efficiency comparable to existing state-of-the-art classifiers. We also discuss how to further increase the stability of our classifier by using it in ensemble model.

Since our GP classifier represents a new technique for one-class classification, there are numerous options to follow when considering future work. One direction could be to investigate various density metrics either in post-hoc analysis or already in the fitness function. Next, one extremely interesting option would be to explore explainability. Here, by explainability we mean being able to understand why GP classifies correctly a specific instance. Even more importantly, if GP does not classify an instance correctly, being able to understand why not and what needs to be added to the model in order to classify that instance correctly.

Acknowledgements This work has been supported in part by Cybersecurity@CSAIL initiative. Additionally, this work has been supported in part by Croatian Science Foundation under the project IP-2014-09-4882.

References

1. IEEE Standard for Floating-Point Arithmetic. IEEE Std 754-2008 pp. 1–70 (2008)
2. Aggarwal, C.C.: Outlier Analysis. Springer Publishing Company, Incorporated (2013)
3. Cao, V.L., Nicolau, M., McDermott, J.: One-Class Classification for Anomaly Detection with Kernel Density Estimation and Genetic Programming. In: Genetic Programming - 19th European Conference, EuroGP 2016, Porto, Portugal, March 30 - April 1, 2016, Proceedings, pp. 3–18 (2016)
4. Chang, C.C., Lin, C.J.: LIBSVM: A library for support vector machines. ACM Transactions on Intelligent Systems and Technology **2**, 27:1–27:27 (2011). Software available at http://www.csie.ntu.edu.tw/~cjlin/libsvm
5. Curry, R., Heywood, M.I.: One-Class Genetic Programming. In: Genetic Programming, 12th European Conference, EuroGP 2009, Tübingen, Germany, April 15–17, 2009, Proceedings, pp. 1–12 (2009)
6. Eddy, W.M.: Defenses Against TCP SYN Flooding Attacks - The Internet Protocol Journal - Volume 9, Number 4 (2017). URL http://www.cisco.com/c/en/us/about/press/internet-protocol-journal/back-issues/table-contents-34/syn-flooding-attacks.html
7. Elsayed, S., Sarker, R., Slay, J.: Evaluating the performance of a differential evolution algorithm in anomaly detection. In: 2015 IEEE Congress on Evolutionary Computation (CEC), pp. 2490–2497 (2015)

8. Folino, G., Pizzuti, C., Spezzano, G.: GP Ensemble for Distributed Intrusion Detection Systems. In: S. Singh, M. Singh, C. Apte, P. Perner (eds.) Pattern Recognition and Data Mining: Third International Conference on Advances in Pattern Recognition, ICAPR 2005, Bath, UK, August 22–25, 2005, Proceedings, Part I, pp. 54–62. Springer Berlin Heidelberg, Berlin, Heidelberg (2005)
9. Habibi, A., et al.: UNB ISCX NSL-KDD DataSet (2017). URL http://nsl.cs.unb.ca/NSL-KDD/
10. Jakobovic, D., et al.: Evolutionary Computation Framework (2016). URL http://ecf.zemris.fer.hr/
11. Khan, S.S., Madden, M.G.: One-Class Classification: Taxonomy of Study and Review of Techniques. CoRR **abs/1312.0049** (2013). URL http://arxiv.org/abs/1312.0049
12. Kuzmanovic, A., Knightly, E.W.: Low-rate tcp-targeted denial of service attacks: the shrew vs. the mice and elephants. In: Proceedings of the 2003 conference on Applications, technologies, architectures, and protocols for computer communications, pp. 75–86. ACM (2003)
13. Ni, X., He, D., Chan, S., Ahmad, F.: Network Anomaly Detection Using Unsupervised Feature Selection and Density Peak Clustering. In: M. Manulis, A.R. Sadeghi, S. Schneider (eds.) Applied Cryptography and Network Security: 14th International Conference, ACNS 2016, Guildford, UK, June 19–22, 2016. Proceedings, pp. 212–227. Springer International Publishing, Cham (2016)
14. Orfila, A., Estevez-Tapiador, J.M., Ribagorda, A.: Evolving High-Speed, Easy-to-Understand Network Intrusion Detection Rules with Genetic Programming.   In: M. Giacobini, A. Brabazon, S. Cagnoni, G.A. Di Caro, A. Ekárt, A.I. Esparcia-Alcázar, M. Farooq, A. Fink, P. Machado (eds.) Applications of Evolutionary Computing: EvoWorkshops 2009: EvoCOMNET, EvoENVIRONMENT, EvoFIN, EvoGAMES, EvoHOT, EvoIASP, EvoINTERACTION, EvoMUSART, EvoNUM, EvoSTOC, EvoTRANSLOG, Tübingen, Germany, April 15–17, 2009. Proceedings, pp. 93–98. Springer Berlin Heidelberg, Berlin, Heidelberg (2009)
15. Overton, M.L.: Numerical Computing with IEEE Floating Point Arithmetic.   Society for Industrial and Applied Mathematics, Philadelphia, PA, USA (2001)
16. Poli, R., Langdon, W.B., McPhee, N.F.: A field guide to genetic programming. Published via http://lulu.com and freely available at http://www.gp-field-guide.org.uk (2008).   (With contributions by J. R. Koza)
17. R Development Core Team: R: A Language and Environment for Statistical Computing.   R Foundation for Statistical Computing, Vienna, Austria (2008). URL http://www.R-project.org. ISBN 3-900051-07-0
18. Schölkopf, B., Platt, J.C., Shawe-Taylor, J.C., Smola, A.J., Williamson, R.C.: Estimating the Support of a High-Dimensional Distribution. Neural Comput. **13**(7), 1443–1471 (2001)
19. Shiravi, A., Shiravi, H., Tavallaee, M., Ghorbani, A.A.: Toward Developing a Systematic Approach to Generate Benchmark Datasets for Intrusion Detection.   Comput. Secur. **31**(3), 357–374 (2012)
20. Song, D., Heywood, M.I., Zincir-Heywood, A.N.: Training genetic programming on half a million patterns: an example from anomaly detection. IEEE Trans. Evolutionary Computation **9**(3), 225–239 (2005)
21. Tavallaee, M., Bagheri, E., Lu, W., Ghorbani, A.A.: A Detailed Analysis of the KDD CUP 99 Data Set.   In: Proceedings of the Second IEEE International Conference on Computational Intelligence for Security and Defense Applications, CISDA'09, pp. 53–58. IEEE Press, Piscataway, NJ, USA (2009)
22. To, C., Elati, M.: A Parallel Genetic Programming for Single Class Classification. In: Proceedings of the 15th Annual Conference Companion on Genetic and Evolutionary Computation, GECCO '13 Companion, pp. 1579–1586. ACM, New York, NY, USA (2013)
23. Tsai, C.F., Hsu, Y.F., Lin, C.Y., Lin, W.Y.: Intrusion detection by machine learning: A review. Expert Systems with Applications **36**(10), 11,994–12,000 (2009)

24. Wang, W., Gombault, S., Guyet, T.: Towards Fast Detecting Intrusions: Using Key Attributes of Network Traffic. In: Proceedings of the 2008 The Third International Conference on Internet Monitoring and Protection, ICIMP '08, pp. 86–91. IEEE Computer Society, Washington, DC, USA (2008)
25. Wu, S.X., Banzhaf, W.: The Use of Computational Intelligence in Intrusion Detection Systems: A Review. Appl. Soft Comput. **10**(1), 1–35 (2010)
26. Zargari, S., Voorhis, D.: Feature Selection in the Corrected KDD-dataset. In: 2012 Third International Conference on Emerging Intelligent Data and Web Technologies, pp. 174–180 (2012)

Chapter 11
Evolution of Space-Partitioning Forest for Anomaly Detection

Zhiruo Zhao, Stuart W. Card, Kishan G. Mehrotra, and Chilukuri K. Mohan

Abstract Previous work proposed a fast one-class anomaly detector using an ensemble of random half-space partitioning trees. The method was shown to be effective and efficient for detecting anomalies in streaming data. However, the parameters were pre-defined, so the random partitions of the data space might not be optimal. Therefore, the aims of this study were to: (a) give some mathematical analysis of the random partitioning trees; and (b) explore optimizing forests for anomaly detection using evolutionary algorithms.

11.1 Anomaly Detection for Streaming Data

The problem of anomaly detection or outlier detection has been well studied in many applications, [4, 7] provide detailed surveys; [1] is a comprehensive reference. While in [5] the outliers/anomalies are defined as "Data objects are grossly different from or inconsistent with the remaining set of data"; in [4], they are defined as "Patterns in data that do not confirm to expected behavior". The definition of what is an anomaly depends on the application. For example, in intrusion detection systems, an anomaly is referred to an intrusion attack; in medical diagnosis, an anomaly could be a patient who is diagnosed of having cancer. Detecting anomalies on static data has been well-studied in [2, 8, 9, 12], and ensemble-based methods were proposed in [3, 13–16]. The focus of this paper is to improve an ensemble method for detecting anomalies in streaming data.

Given a data stream arriving at continuous time stamps $T_1, T_2, \ldots, T_n, \ldots$, each data X_k is a high-dimensional data point containing d attributes, denoted by $X_k = (x_k^1, x_k^2, \ldots, x_k^d, label)$, where $label \in \{normal, abnormal\}$ is unknown a priori. The problem of finding anomalies from streaming data is to separate the data points with $label = abnormal$ from the majority data points with $label = normal$. We

Z. Zhao (✉) · S. W. Card · K. G. Mehrotra · C. K. Mohan
Syracuse University, Syracuse, NY, USA
e-mail: zzhao11@syr.edu; stu.card@critical.com; mehrotra@syr.edu; mohan@syr.edu

© Springer International Publishing AG, part of Springer Nature 2018
W. Banzhaf et al. (eds.), *Genetic Programming Theory and Practice XV*,
Genetic and Evolutionary Computation,
https://doi.org/10.1007/978-3-319-90512-9_11

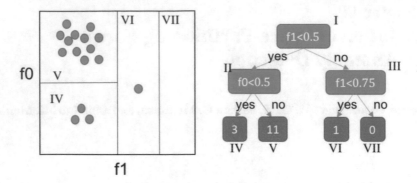

Fig. 11.1 An illustration of dataspace partition by one HSTree

use A to denote the set of attributes, namely, $A = \{a_i | i = 1, \ldots, d\}$. By definition of anomaly detection, abnormal points are far fewer than normal points. With absence of labeled training data, to find anomalies is to find a set of rules which can separate the minor data points from major data points and denote the minority points as anomalies.

Previous work in Half-Space Trees [11] adopt an ensemble method which combines results from a set of full binary trees where each tree node selects a random attribute from A and partitions that attribute in half. An illustration can be found in Fig. 11.1. In their method, each tree node is associated with a mass profile which records the number of points falling in that partition. The method constructs t trees without using any data; to get the mass profile for each node, they use the first ϕ data points and assume they are all normal points. The intuition behind this method is that anomalies fall in sparse partitions while normal objects fall in dense areas. In order to deal with streaming data, the method retains the mass profile every ϕ data points, records the newly arrived points in their *latest* window, and stores the previous batch in the *reference* window. They use the mass profile in *reference* window to detect anomalies, let $Node$ represents the node a point falls in (reaches the maximum height or contains less than $sizeLimit$ points), and, $Node.r$ is the mass profile in *reference* window, $Node.height$ denotes the node height, $Score(x, T)$ is the anomaly score for object x in tree T, $Score(x, T) = Node.r * 2^{Node.height}$. This score will be equivalent for different partitions at different tree levels if the data distribution is uniform. For rare anomalies, the score is far less than the score of normal objects.

This method has shown its advance in detecting anomalies in low-dimensional data streams, however, the previous paper suffers from the following aspects:

- the tree height and number trees are selected using pre-defined values without any mathematical justification w.r.t dimensionality of dataset
- each tree partitions the data spaces randomly, building random trees in non-informative attributes could lead to pollution of final results

In this paper, we address the aforementioned problems:

- we give a mathematical justification of required tree height and number of trees based on the dimensionality of the dataset by casting the problem as a classical coupon collector problem
- we better partition the data space for anomaly detection by using an EA to maximize the likelihood of separating outliers from normal observations

11.2   Analysis of Random Trees

11.2.1   Number of Nodes Needed to Capture Anomaly Characteristics

For our purposes, some features may be merely noise, i.e. uninformative for anomaly detection. When such features are present, more nodes may be needed to capture anomaly characteristics.

11.2.1.1   Theoretical Justification

Let $m = |A|$ be total number of features, $k = |I|$ be the number of informative features, so $m - k = |A - I|$ is the number of noise features that are not contributing to anomaly detection. In constructing a random forest, if each node is randomly selected from m features, and $Pr(SelectInform)$ represents the probability that informative features are selected at each node, then, it follows:

$$Pr(SelectInform) = k/m.$$

For each random tree t, let h be the height of t, and $nodeCount(t) = 2^h - 1$ since t is a complete binary tree. Let $IF(t)$ be the number of informative features covered by tree t and follows a binomial distribution,

$$IF(t) \sim B(n, p), n = nodeCount(t), p = \frac{k}{m}.$$

When the number of noisy feature increases, i.e. $\frac{k}{m}$ is decreasing, we need more nodes to cover more informative features. If the height of trees are the same, then, this is equivalent to say that we will need more trees in our forest to cover a substantial fraction of the informative features.

11.2.1.2   Experimental Results

We construct two synthetic datasets to address this problem. We choose two datasets which are used in the original HSTree paper [11] where their method was shown to be very effective. **Http dataset** contains 567,497 data points, three features and 0.4% anomalies presented. **CoverType dataset** contains 286,048 data points, 10 features and 0.9% anomalies presented. We add spurious features drawn from a uniform distribution to the two data sets, to simulate the scenario where anomalies may be hidden by such noise features. We denote the data with noisy features **Http dataset** as **syn-1** and **CoverType dataset** as **syn-2**. In the following we use area under the ROC curve (AUC) [6] score to evaluate the performance of our algorithm. A perfect detector will have an AUC score of 1.0 while a random one gives 0.5.

We designed and performed three sets of experiments. In the first experiment we compare the performance with increased noisy features. In the second experiment we insert 40 noisy features and compare how increased number of trees affect model performance. In the third we test how the height of trees affect performance when dimensions are different.

In Fig. 11.2 we show the results. From these figures, we observe that:

- random forests need more trees to cover all the features
- random forests need higher trees to 'reach' the feature values

showing the experimental results are consistent with our theoretical justification.

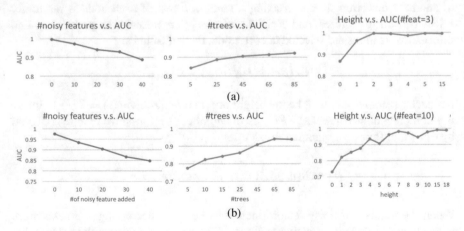

Fig. 11.2 Performance comparison in the presence of noise features. (**a**) Performance with syn-1. (**b**) Performance with syn-2

11.2.2   Number and Height of Trees

Given the demonstration in the previous section that more trees are needed to capture anomaly characteristics in the presence of noise features, in this section we calculate the expected number and height of trees required when the number of features is fixed.

11.2.2.1   The Coupon Collector's Problem: Analysis of Tree Height

In the coupon collector's problem [10], there are d types of coupons and they are drawn at random at each trial. Let r be the number of trials for one to collect at least one copy of each of the d types of coupons. The goal of the coupon collector's problem is to find out what is the relationship between r and d.

The similarity between the random trees and the coupon collector's problem is that if we treat each feature as a type of coupon, each detection path in the tree can be treated as an experiment with n trials—where n is the number of nodes in a random tree of height h. If an anomaly is jointly described by d features, each tree should capture at least one copy of each of the d features. To study the relationship between the number of nodes n and number of features present, we adopt the theoretical results for the coupon collector's problem.

We show that when the tree height is

$$h = \log_2(\beta d \ln d + 1),$$

the probability that at least one of the features is not captured is bounded by $d^{-(\beta-1)}$, where $\beta > 1$.

Let X_d be a random variable defined to be the number of nodes required to collect at least one copy of each type of the d features. The expected number of nodes is

$$E[X_d] = d \sum_{i=1}^{d} \frac{1}{i} = d H_d$$

where H_d is the harmonic sum [10].

Let σ_i^n be the event that feature i is not selected in the n nodes, the probability of this event is:

$$Pr[\sigma_i^n] = (1 - \frac{1}{d})^n \le e^{-\frac{n}{d}},$$

for $n = \beta d \ln d$, this bound is $d^{-\beta}$, where $\beta > 1$ is a constant.

Thus, the probability that at least one of the features is not captured in the n nodes is

$$Pr[\cup_{i=1}^{d}\sigma_i^n] \leq \sum_{i=1}^{d} Pr[\sigma_i^n] \leq \sum_{i=1}^{d} d^{-\beta} = d^{-(\beta-1)},$$

for a random tree with number of nodes $n = \beta d \ln d$, consequently, the tree height $h = \log_2(n+1) = \log_2(\beta d \ln d + 1)$.

11.2.2.2 Number of Trees T for a Given Tree Height h and Number of Features d

Given tree height h, each tree has $n = 2^h - 1$ nodes. For T such trees, the total number of nodes is nT. Number of trees T is chosen such that the probability that each feature occurs at least in one of the T trees should be larger than $1 - v$. From the results from the coupon collector problem, we have:

$$Pr(X_d = k) = \sum_{j=0}^{d-1}(-1)^j \binom{d-1}{j}(1 - \frac{1+j}{d})^{k-1}.$$

It is desired that:

$$1 - (1 - Pr[d \leq X_d \leq nT]) \geq 1 - v \tag{11.1}$$

$$Pr[d \leq X_d \leq nT] \geq 1 - v \tag{11.2}$$

$$\sum_{i=d}^{nT} Pr(X_d = i) \geq 1 - v \tag{11.3}$$

$$\sum_{i=d}^{nT}\sum_{j=0}^{d-1}(-1)^j \binom{d-1}{j}\left(1 - \frac{1+j}{d}\right)^{k-1} \geq 1 - v \tag{11.4}$$

This is a combinatorial problem, and numerical solutions are shown in Fig. 11.3.

11.3 Use EA to Better Partition Data Space

Currently, in our model, when we build detection trees, each tree denotes a random partition. In each partition, attributes are selected randomly and then split into half. We want to further improve the partition generation process given that picking up irrelevant features might dilute the general detection performance, and instead of splitting the attribute into half space, we want to find a better split point where extreme values (anomalies) can be better separated from the normal data.

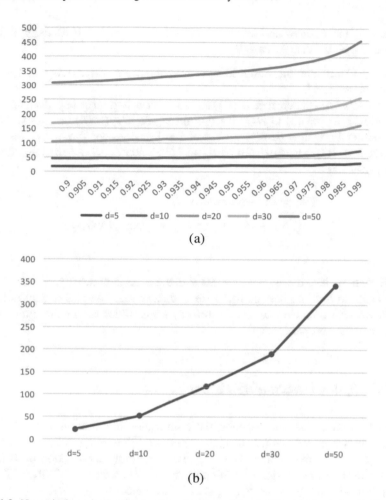

Fig. 11.3 Numerical results for the number of trees and tree height. (**a**) Y-axis is nT, X-axis is $1 - \nu$. (**b**) Y-axis is nT, X-axis is number of dimensions

11.3.1 How to Partition the Data Space to Separate Outliers

Though the definition of outliers/anomalies depends on the application, a general heuristic is that outliers reside in sparser regions of the data space than do normal observations. For a dataset D, we want to find a *partition* that best separates the anomalies from the normal objects. Thus, we define:

$$Density(partition) = \sum_{i=1}^{P} density(partition_i)$$

Claim All the outliers are separated in a data space DS if Density d is the maximum over the data space DS.

Proof by Contradiction Suppose d is maximum and there exists at least one outlier not separated.

$\exists o \in Outliers$ such that o is grouped with a normal group NG and form a partition $NG' = \{o\} \cup NG$. By definition, $density(NG') < density(NG)$ since outlier lie in low density area. Let the Density over $DS \setminus NG'$ be $d_0 = d - density(NG)$. There exists a partition $\{DS \setminus NG', o, NG\}$ such that the total density over it is

$$
\begin{aligned}
d_1 &= d_0 + density(NG) \\
&= d - density(NG') + density(NG) \\
&> d
\end{aligned}
$$

which contradicts to the assumption that d is the maximum density.

In order to find the best partition for a dataset that contains n data points, it requires $O(2^n)$ time complexity. Therefore, we consider EA as the optimization tool to find the best partitions.

11.3.2   Space-Partitioning Forest

Each detection tree should partition the data space such that normal objects are grouped in denser and outliers in sparser regions. Attempting to find one optimal tree may result in over-fitting. Thus we aim at building a collection of T space-partitioning trees, i.e. a space-partitioning forest. Therefore, for T trees of same height h full-binary trees, the query for each data object is $O(T \cdot h)$.

11.3.3   Individual Representation

The goal is to find a collection of trees to better capture anomalies. Each individual is a collection of T trees, and each tree is represented in its *level-order* traversal representation. Each tree of height h (start from 1) consists of $2^h - 1$ interior nodes, each interior node is represented as a tuple: *(attId, splitVal)*, represents the ID of the attribute and the cutoff value at that node. Thus, each tree of height h is represented as a vector of nodes:

$$
< node_1, node_2, \ldots, node_{2^h-1} >
$$

Each individual is a collection of T trees, represented as a set of T such vectors.

11.3.4 Cost Function

In the previous section, we defined a density function for a partition recursively as the sum of the densities of its constituent partitions, but did not address the base case. Here, to maximize density, we approximate its inverse (sparsity) by the maximum distance between points in a sub-partition (node):

$$MaxDist(node) = \text{Max}\{distance(centroid(node), node.data_i),$$

$$i = 1 \ldots |node.data|\}$$

Cost function of each tree is defined as the averaged maximum distances among all its leaf nodes:

$$cost(tree) = \frac{1}{2^H} \sum_{l=1}^{2^H} \text{MaxDist}(leaf_l).$$

Each individual in EA is defined as a collection of T trees. The cost of individual is defined as the average of the costs of the T trees:

$$cost(individual) = \frac{1}{T} \sum_{i=1}^{T} cost(tree_i)$$

The goal is to minimize the cost.

We first want to examine the effectiveness of our cost function in finding anomalies. The following figures show two partitions created from two different detection trees.

We notice in Fig. 11.4 that when cost is high, anomalies cannot be separated using the partition generated. For example, the two trees built in Fig. 11.4 partition in total 101 points, 1 outlier is located at (0,0) and 100 normal objects form two clusters. For the tree in Fig. 11.4a, each tree leaf contains 15, 50, 34 and 2 points respectively; the partitioned data space is shown in the left, outlier is partitioned with some of the normal objects from the lower right corner, and the cost for this tree is equal to $(2.6 + 2.5 + 3.7 + 0.1)/4 = 2.225$. While the tree built in Fig. 11.4b partitions the outlier from the normal observations, and the normal objects are grouped densely in other partitions, and this tree has a cost of $(2.6+2.5)/4 = 1.275$, which is lower than before. This illustrates the effectiveness of our cost function in separating anomalies from normal objects (Fig. 11.5).

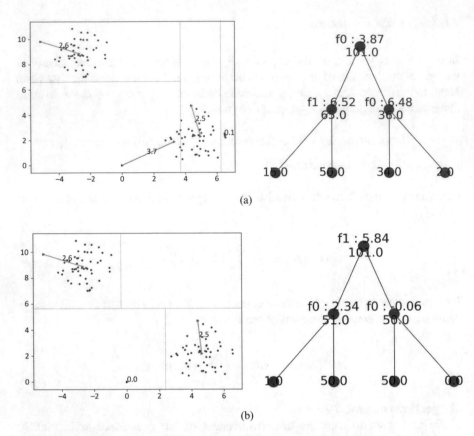

(a)

(b)

Fig. 11.4 Cost computation for two trees with different degrees of separation of outliers from the other data points. (**a**) Cost is higher when outlier is partitioned with normal objects. (**b**) Cost is lower when outlier is separated

Input: mutation rate *pM*, individual *individual*, current generation *gen*
Output: return a mutated individual
counter = 0 ;
prevCost = cost(*individual*) ;
mute = randomly change one node from a random tree in *individual* ;
while *counter < N and cost(mute) >= prevCost* **do**
  | *counter* + + ;
  | mute = randomly change one node from a random tree in *individual* ;
  | prevCost = cost(*individual*) ;
end

Fig. 11.5 Mutation algorithm

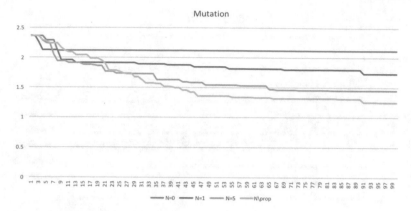

Fig. 11.6 Mutation with different strategies for synthetic data

11.3.5 *Mutation*

The idea of mutation is to not change the individual drastically, instead we modify
one (or few) nodes in one tree to keep the diversity. In order to find better solution,
we discard the offsprings which give us higher cost. Intuitionally, this will lead to a
hill-climbing searching procedure which could be unnecessary and computationally
expensive, therefore, we add some contraints in our mutation procedure such that
we set a counter and constrain it to be less than N times to find a better mutate. In
our experiment, we tried $N = 0, 1, 5$ and $N \propto$ current generation. When $N = 0$,
that means the constraint is removed. When $N \propto$ current generation used, it means
we want finer tuning at the end of convergence. The results for some synthetic data
arc shown in Fig. 11.6, where we observe that using the last strategy has higher cost
reducing rate than the others.

11.3.6 *Crossover*

In our algorithm, each individual is a set of independent detection trees. For
crossover, we apply single-point crossover on the two parents sets. For example,
for two individual $\{T_1, T_2\}$ and $\{T_3, T_4\}$, after applying single point crossover at the
midpoint, we obtained two offspring $\{T_1, T_4\}$ and $\{T_2, T_3\}$.

   The cost over iterations for different crossover probability pC is shown in
Fig. 11.7.

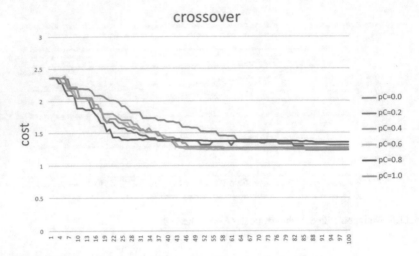

Fig. 11.7 Different crossover probability for synthetic data

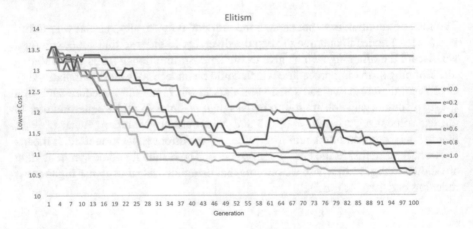

Fig. 11.8 Elitism with different *e* for synthetic data

11.3.7   Selection

We add elitism in our selection procedure. Which means in each iteration, we keep
$e * 100\%$ elites from the sorted population in the next iteration. For reproductive
selection, we choose parents with probability inversely related to their cost ('fitness
proportionate' selection).

The cost over iterations for different *e* is shown in Fig. 11.8, Moderate values of
e (from 0.2 to 0.6) gave better results than extremely strong elitism or no elitism.

Input: population size N, mutation probability pM, crossover probability pC
Output: return the possible best individual
Pop = generate N random initial individuals;
$isTerminated$ = false;
$best$ = None ;
while *not isTerminated* **do**
$\quad C = \{\text{cost}(i)|\forall i \in Pop\}$;
$\quad Pop' = \emptyset$;
$\quad$ **while** $|Pop'| < n$ **do**
$\quad\quad parents = \text{select}(Pop, C)$;
$\quad\quad offspring = \text{reproduce}(parents, pM, pC)$;
$\quad\quad$ add $offspring$ to Pop';
$\quad$ **end**
$\quad Pop = Pop'$;
$\quad isTerminate, best = \text{testTermination}(Pop, best)$;
end
return *best*;

Fig. 11.9   Overall evolutionary algorithm

Input: a population Pop, the best individual *best* previously seen
Output: a tuple (*isTerminated, individual*) where *isTerminated* is True if terminate,
$\quad\quad\quad$ *individual* is the best individual returned
$individual = \text{argmin}_i\{\text{cost}(i)|\forall i \in Pop\}$;
if *reach maximum generation* **then**
$\quad$ **return** (True, *individual*);
end
else
$\quad$ **if** $|cost(individual) - cost(best)| < \delta$ **then**
$\quad\quad$ **return** (True, *individual*);
$\quad$ **end**
$\quad$ **else**
$\quad\quad$ **return** (False, *individual*);
$\quad$ **end**
end

Fig. 11.10   TestTermination algorithm

11.3.8   Algorithms

The overall EA is sketched as Fig. 11.9 and its termination test as Fig. 11.10.

11.3.9   Preliminary Results for EA

In this section, we show some preliminary results for comparison of using EA to
generate the random trees with pure random trees generation. We generate synthetic

Fig. 11.11 Synthetic data 1.
(**a**) Synthetic data—normal
data from gaussian, outliers
uniformly distributed. (**b**)
Cost vs AUC for using EA to
generate random trees

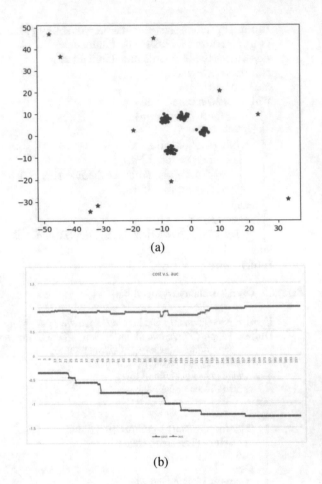

(a)

(b)

dataset where the normal data comes from Gaussian distribution while outliers are uniformly distributed, the data is shown in Fig. 11.11. Results for using this simple EA is shown in Fig. 11.11. We observe that EA successfully finds all the outliers (AUC score is 1.0) when it converges.

Another synthetic dataset is shown in Fig. 11.12. In this experiment, we want to see whether the EA can better separate the outliers between clusters better than purely random trees. We fix the tree height at 4 for both purely random and EA-generated trees. We observe that when using 10 trees, the AUC for random is 0.64 while EA is 0.94. Notice that in this experiment, when we use large number of trees of large height, the improvement of using EA over random tree is not very significant. The reason is if given enough tree cost (i.e. tree height and number of trees), the probability of covering all combinations of all features is high (refer to our previous mathematical analysis.)

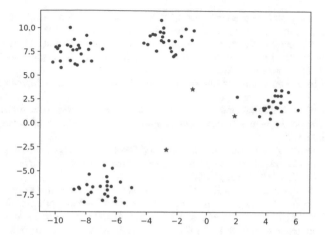

Fig. 11.12 Synthetic data 2–4 clusters, outliers are inserted in between

11.4 Conclusion and Future Work

In this study, we first calculate the height and number of trees expected to be needed by treating the random partitioning tree problem as a coupon collector problem. Secondly, we formulate an optimization problem for separating anomalies from normal objects, and demonstrate performance improvement on synthetic datasets using a simple EA to build optimized rather than purely random trees. The trade-off between the performance gain and computational cost of using an EA for this application remains to be addressed. Rewarding diversity should be considered in future work. Finally, this approach should be attempted on a variety of real world datasets.

References

1. Charu C Aggarwal. *Outlier analysis*. Springer Science & Business Media, 2013.
2. Markus M Breunig, Hans-Peter Kriegel, Raymond T Ng, and Jörg Sander. Lof: identifying density-based local outliers. In *ACM Sigmod Record*, volume 29, pages 93–104. ACM, 2000.
3. João BD Cabrera, Carlos Gutiérrez, and Raman K Mehra. Ensemble methods for anomaly detection and distributed intrusion detection in mobile ad-hoc networks. *Information Fusion*, 9(1):96–119, 2008.
4. Varun Chandola, Arindam Banerjee, and Vipin Kumar. Anomaly detection: A survey. *ACM computing surveys (CSUR)*, 41(3):15, 2009.
5. Jiawei Han, Jian Pei, and Micheline Kamber. *Data mining: concepts and techniques*. Elsevier, 2011.
6. James A Hanley and Barbara J McNeil. The meaning and use of the area under a receiver operating characteristic (roc) curve. *Radiology*, 143(1):29–36, 1982.
7. Victoria J Hodge and Jim Austin. A survey of outlier detection methodologies. *Artificial Intelligence Review*, 22(2):85–126, 2004.

 8. Huaming Huang, Kishan Mehrotra, and Chilukuri K Mohan. Rank-based outlier detection. *Journal of Statistical Computation and Simulation*, 83(3):518–531, 2013.
 9. Wen Jin, Anthony KH Tung, Jiawei Han, and Wei Wang. Ranking outliers using symmetric neighborhood relationship. In *Advances in Knowledge Discovery and Data Mining*, pages 577–593. Springer, 2006.
10. Rajeev Motwani and Prabhakar Raghavan. *Randomized algorithms*. Chapman & Hall/CRC, 2010.
11. Swee Chuan Tan, Kai Ming Ting, and Tony Fei Liu. Fast anomaly detection for streaming data. In *IJCAI Proceedings-International Joint Conference on Artificial Intelligence*, volume 22, page 1511, 2011.
12. Jian Tang, Zhixiang Chen, Ada Wai-Chee Fu, and David W Cheung. Enhancing effectiveness of outlier detections for low density patterns. In *Advances in Knowledge Discovery and Data Mining*, pages 535–548. Springer, 2002.
13. Lena Tenenboim-Chekina, Lior Rokach, and Bracha Shapira. Ensemble of feature chains for anomaly detection. In *Multiple Classifier Systems*, pages 295–306. Springer, 2013.
14. Zhiruo Zhao, Kishan G Mehrotra, and Chilukuri K Mohan. Ensemble algorithms for unsupervised anomaly detection. In *Current Approaches in Applied Artificial Intelligence*, pages 514–525. Springer, 2015.
15. Zhiruo Zhao, Chilukuri K. Mohan, and Kishan G. Mehrotra. Adaptive sampling and learning for unsupervised outlier detection. In *Proceedings of the Twenty-Ninth International Florida Artificial Intelligence Research Society Conference, FLAIRS 2016, Key Largo, Florida, May 16–18, 2016*, pages 460–466, 2016.
16. Arthur Zimek, Matthew Gaudet, Ricardo JGB Campello, and Jörg Sander. Subsampling for efficient and effective unsupervised outlier detection ensembles. In *Proceedings of the 19th ACM SIGKDD international conference on Knowledge discovery and data mining*, pages 428–436. ACM, 2013.

Index

© Springer International Publishing AG, part of Springer Nature 2018
W. Banzhaf et al. (eds.), *Genetic Programming Theory and Practice XV*,
Genetic and Evolutionary Computation, https://doi.org/10.1007/978-3-319-90512-9

Printed in the United States
By Bookmasters